The Application of Hidden Markov Models in Speech Recognition

The Application of Hidden Markov Models in Speech Recognition

Mark Gales

Cambridge University Engineering Department
Cambridge
CB2 1PZ
UK
mjfg@eng.cam.ac.uk

Steve Young

Cambridge University Engineering Department
Cambridge
CB2 1PZ
UK
sjy@eng.cam.ac.uk

Boston – Delft

Foundations and Trends$^{\circledR}$ in Signal Processing

Published, sold and distributed by:
now Publishers Inc.
PO Box 1024
Hanover, MA 02339
USA
Tel. +1-781-985-4510
www.nowpublishers.com
sales@nowpublishers.com

Outside North America:
now Publishers Inc.
PO Box 179
2600 AD Delft
The Netherlands
Tel. +31-6-51115274

The preferred citation for this publication is M. Gales and S. Young, The Application of Hidden Markov Models in Speech Recognition, Foundations and Trends$^{\circledR}$ in Signal Processing, vol 1, no 3, pp 195–304, 2007

ISBN: 978-1-60198-120-2
© 2008 M. Gales and S. Young

Foundations and Trends® in Signal Processing

Volume 1 Issue 3, 2007

Editorial Board

Editorial Scope

Foundations and Trends® in Signal Processing will publish survey and tutorial articles on the foundations, algorithms, methods, and applications of signal processing including the following topics:

- Adaptive signal processing
- Audio signal processing
- Biological and biomedical signal processing
- Complexity in signal processing
- Digital and multirate signal processing
- Distributed and network signal processing
- Image and video processing
- Linear and nonlinear filtering
- Multidimensional signal processing
- Multimodal signal processing
- Multiresolution signal processing
- Nonlinear signal processing
- Randomized algorithms in signal processing
- Sensor and multiple source signal processing, source separation
- Signal decompositions, subband and transform methods, sparse representations
- Signal processing for communications
- Signal processing for security and forensic analysis, biometric signal processing
- Signal quantization, sampling, analog-to-digital conversion, coding and compression
- Signal reconstruction, digital-to-analog conversion, enhancement, decoding and inverse problems
- Speech/audio/image/video compression
- Speech and spoken language processing
- Statistical/machine learning
- Statistical signal processing
- Classification and detection
- Estimation and regression
- Tree-structured methods

Information for Librarians

Foundations and Trends® in Signal Processing, 2007, Volume 1, 4 issues. ISSN paper version 1932-8346. ISSN online version 1932-8354. Also available as a combined paper and online subscription.

Foundations and Trends® in
Signal Processing
Vol. 1, No. 3 (2007) 195–304
© 2008 M. Gales and S. Young
DOI: 10.1561/2000000004

The Application of Hidden Markov Models in Speech Recognition

Mark Gales[1] and Steve Young[2]

[1] *Cambridge University Engineering Department, Trumpington Street, Cambridge, CB2 1PZ, UK, mjfg@eng.cam.ac.uk*
[2] *Cambridge University Engineering Department, Trumpington Street, Cambridge, CB2 1PZ, UK, sjy@eng.cam.ac.uk*

Abstract

Hidden Markov Models (HMMs) provide a simple and effective framework for modelling time-varying spectral vector sequences. As a consequence, almost all present day large vocabulary continuous speech recognition (LVCSR) systems are based on HMMs.

Whereas the basic principles underlying HMM-based LVCSR are rather straightforward, the approximations and simplifying assumptions involved in a direct implementation of these principles would result in a system which has poor accuracy and unacceptable sensitivity to changes in operating environment. Thus, the practical application of HMMs in modern systems involves considerable sophistication.

The aim of this review is first to present the core architecture of a HMM-based LVCSR system and then describe the various refinements which are needed to achieve state-of-the-art performance. These refinements include feature projection, improved covariance modelling, discriminative parameter estimation, adaptation and normalisation, noise compensation and multi-pass system combination. The review concludes with a case study of LVCSR for Broadcast News and Conversation transcription in order to illustrate the techniques described.

Contents

1

Introduction

Automatic continuous speech recognition (CSR) has many potential applications including command and control, dictation, transcription of recorded speech, searching audio documents and interactive spoken dialogues. The core of all speech recognition systems consists of a set of statistical models representing the various sounds of the language to be recognised. Since speech has temporal structure and can be encoded as a sequence of spectral vectors spanning the audio frequency range, the hidden Markov model (HMM) provides a natural framework for constructing such models [13].

HMMs lie at the heart of virtually all modern speech recognition systems and although the basic framework has not changed significantly in the last decade or more, the detailed modelling techniques developed within this framework have evolved to a state of considerable sophistication (e.g. [40, 117, 163]). The result has been steady and significant progress and it is the aim of this review to describe the main techniques by which this has been achieved.

The foundations of modern HMM-based continuous speech recognition technology were laid down in the 1970's by groups at Carnegie-Mellon and IBM who introduced the use of discrete density HMMs

[11, 77, 108], and then later at Bell Labs [80, 81, 99] where continuous density HMMs were introduced.[1] An excellent tutorial covering the basic HMM technologies developed in this period is given in [141].

Reflecting the computational power of the time, initial development in the 1980's focussed on either discrete word speaker dependent large vocabulary systems (e.g. [78]) or whole word small vocabulary speaker independent applications (e.g. [142]). In the early 90's, attention switched to continuous speaker-independent recognition. Starting with the artificial 1000 word *Resource Management* task [140], the technology developed rapidly and by the mid-1990's, reasonable accuracy was being achieved for unrestricted speaker independent dictation. Much of this development was driven by a series of DARPA and NSA programmes [188] which set ever more challenging tasks culminating most recently in systems for multilingual transcription of broadcast news programmes [134] and for spontaneous telephone conversations [62].

Many research groups have contributed to this progress, and each will typically have its own architectural perspective. For the sake of logical coherence, the presentation given here is somewhat biassed towards the architecture developed at Cambridge University and supported by the HTK Software Toolkit [189].[2]

The review is organised as follows. Firstly, in *Architecture of a HMM-Based Recogniser* the key architectural ideas of a typical HMM-based recogniser are described. The intention here is to present an overall system design using very basic acoustic models. In particular, simple single Gaussian diagonal covariance HMMs are assumed. The following section *HMM Structure Refinements* then describes the various ways in which the limitations of these basic HMMs can be overcome, for example by transforming features and using more complex HMM output distributions. A key benefit of the statistical approach to speech recognition is that the required models are trained automatically on data.

---

[1] This very brief historical perspective is far from complete and out of necessity omits many other important contributions to the early years of HMM-based speech recognition.

[2] Available for free download at `htk.eng.cam.ac.uk`. This includes a recipe for building a state-of-the-art recogniser for the Resource Management task which illustrates a number of the approaches described in this review.

The section *Parameter Estimation* discusses the different objective functions that can be optimised in training and their effects on performance. Any system designed to work reliably in real-world applications must be robust to changes in speaker and the environment. The section on *Adaptation and Normalisation* presents a variety of generic techniques for achieving robustness. The following section *Noise Robustness* then discusses more specialised techniques for specifically handling additive and convolutional noise. The section *Multi-Pass Recognition Architectures* returns to the topic of the overall system architecture and explains how multiple passes over the speech signal using different model combinations can be exploited to further improve performance. This final section also describes some actual systems built for transcribing English, Mandarin and Arabic in order to illustrate the various techniques discussed in the review. The review concludes in *Conclusions* with some general observations and conclusions.

2

Architecture of an HMM-Based Recogniser

The principal components of a large vocabulary continuous speech recogniser are illustrated in Figure 2.1. The input audio waveform from a microphone is converted into a sequence of fixed size acoustic vectors $Y_{1:T} = y_1, \ldots, y_T$ in a process called feature extraction. The decoder then attempts to find the sequence of words $w_{1:L} = w_1, \ldots, w_L$ which is most likely to have generated Y, i.e. the decoder tries to find

$$\hat{w} = \arg \max_{w} \{ P(w|Y) \}. \tag{2.1}$$

However, since $P(w|Y)$ is difficult to model directly,[1] Bayes' Rule is used to transform (2.1) into the equivalent problem of finding:

$$\hat{w} = \arg \max_{w} \{ p(Y|w) P(w) \}. \tag{2.2}$$

The likelihood $p(Y|w)$ is determined by an *acoustic model* and the prior $P(w)$ is determined by a *language model*.[2] The basic unit of sound

---

[1] There are some systems that are based on discriminative models [54] where $P(w|Y)$ is modelled directly, rather than using generative models, such as HMMs, where the observation sequence is modelled, $p(Y|w)$.

[2] In practice, the acoustic model is not normalised and the language model is often scaled by an empirically determined constant and a word insertion penalty is added i.e., in the log domain the total likelihood is calculated as $\log p(Y|w) + \alpha \log(P(w)) + \beta|w|$ where α is typically in the range 8–20 and β is typically in the range $0--20$.

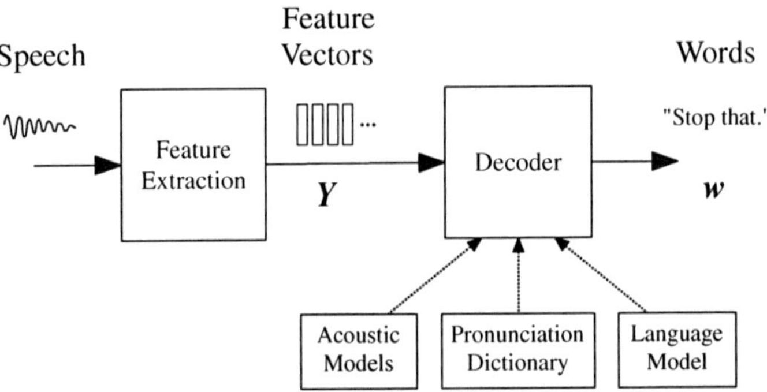

Fig. 2.1 Architecture of a HMM-based Recogniser.

represented by the acoustic model is the *phone*. For example, the word
"bat" is composed of three phones /b/ /ae/ /t/. About 40 such phones
are required for English.

For any given w, the corresponding acoustic model is synthesised by concatenating phone models to make words as defined by
a pronunciation dictionary. The parameters of these phone models
are estimated from training data consisting of speech waveforms and
their orthographic transcriptions. The language model is typically
an N-gram model in which the probability of each word is condi-
tioned only on its $N - 1$ predecessors. The N-gram parameters are
estimated by counting N-tuples in appropriate text corpora. The
decoder operates by searching through all possible word sequences
using pruning to remove unlikely hypotheses thereby keeping the search
tractable. When the end of the utterance is reached, the most likely
word sequence is output. Alternatively, modern decoders can generate lattices containing a compact representation of the most likely
hypotheses.

The following sections describe these processes and components in
more detail.

2.1 Feature Extraction

The feature extraction stage seeks to provide a compact representation of the speech waveform. This form should minimise the loss of

information that discriminates between words, and provide a good match with the distributional assumptions made by the acoustic models. For example, if diagonal covariance Gaussian distributions are used for the state-output distributions then the features should be designed to be Gaussian and uncorrelated.

Feature vectors are typically computed every 10 ms using an overlapping analysis window of around 25 ms. One of the simplest and most widely used encoding schemes is based on *mel-frequency cepstral coefficients (MFCCs)* [32]. These are generated by applying a truncated discrete cosine transformation (DCT) to a log spectral estimate computed by smoothing an FFT with around 20 frequency bins distributed non-linearly across the speech spectrum. The non-linear frequency scale used is called a *mel scale* and it approximates the response of the human ear. The DCT is applied in order to smooth the spectral estimate and approximately decorrelate the feature elements. After the cosine transform the first element represents the average of the log-energy of the frequency bins. This is sometimes replaced by the log-energy of the frame, or removed completely.

Further psychoacoustic constraints are incorporated into a related encoding called *perceptual linear prediction (PLP)* [74]. PLP computes linear prediction coefficients from a perceptually weighted non-linearly compressed power spectrum and then transforms the linear prediction coefficients to cepstral coefficients. In practice, PLP can give small improvements over MFCCs, especially in noisy environments and hence it is the preferred encoding for many systems [185].

In addition to the spectral coefficients, first order (delta) and second-order (delta–delta) regression coefficients are often appended in a heuristic attempt to compensate for the conditional independence assumption made by the HMM-based acoustic models [47]. If the original (static) feature vector is $\boldsymbol{y}_t^{\mathrm{s}}$, then the delta parameter, $\Delta\boldsymbol{y}_t^{\mathrm{s}}$, is given by

$$\Delta\boldsymbol{y}_t^{\mathrm{s}} = \frac{\sum_{i=1}^n w_i \left(\boldsymbol{y}_{t+i}^{\mathrm{s}} - \boldsymbol{y}_{t-i}^{\mathrm{s}}\right)}{2\sum_{i=1}^n w_i^2} \tag{2.3}$$

where n is the window width and w_i are the regression coefficients.[3] The delta–delta parameters, $\Delta^2 \boldsymbol{y}_t^{\mathsf{s}}$, are derived in the same fashion, but using differences of the delta parameters. When concatenated together these form the feature vector $\boldsymbol{y}_t$,

$$\boldsymbol{y}_t = \left[\boldsymbol{y}_t^{\mathsf{s}\mathsf{T}} \quad \Delta \boldsymbol{y}_t^{\mathsf{s}\mathsf{T}} \quad \Delta^2 \boldsymbol{y}_t^{\mathsf{s}\mathsf{T}} \right]^{\mathsf{T}}. \tag{2.4}$$

The final result is a feature vector whose dimensionality is typically around 40 and which has been partially but not fully decorrelated.

2.2 HMM Acoustic Models (Basic-Single Component)

As noted above, each spoken word w is decomposed into a sequence of K_w basic sounds called *base phones*. This sequence is called its pronunciation $\boldsymbol{q}_{1:K_w}^{(w)} = q_1, \ldots, q_{K_w}$. To allow for the possibility of multiple pronunciations, the likelihood $p(\boldsymbol{Y}|w)$ can be computed over multiple pronunciations[4]

$$p(\boldsymbol{Y}|w) = \sum_Q p(\boldsymbol{Y}|Q)P(Q|w), \tag{2.5}$$

where the summation is over all valid pronunciation sequences for $\boldsymbol{w}$, Q is a particular sequence of pronunciations,

$$P(Q|w) = \prod_{l=1}^{L} P(\boldsymbol{q}^{(w_l)}|w_l), \tag{2.6}$$

and where each $\boldsymbol{q}^{(w_l)}$ is a valid pronunciation for word w_l. In practice, there will only be a very small number of alternative pronunciations for each w_l making the summation in (2.5) easily tractable.

Each base phone q is represented by a continuous density HMM of the form illustrated in Figure 2.2 with transition probability parameters $\{a_{ij}\}$ and output observation distributions $\{b_j()\}$. In operation, an HMM makes a transition from its current state to one of its connected states every time step. The probability of making a particular

---

[3] In HTK to ensure that the same number of frames is maintained after adding delta and delta–delta parameters, the start and end elements are replicated to fill the regression window.

[4] Recognisers often approximate this by a *max* operation so that alternative pronunciations can be decoded as though they were alternative word hypotheses.

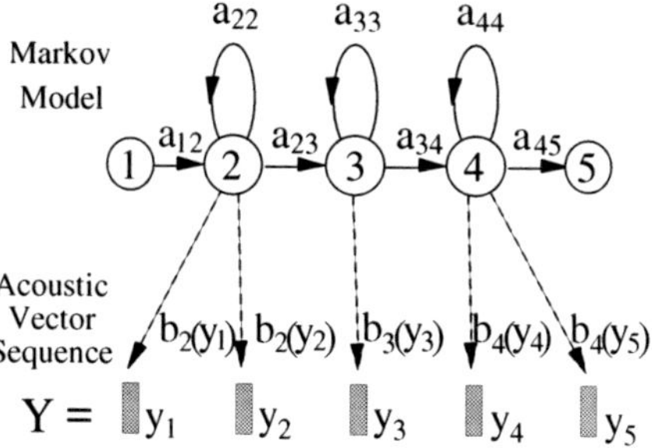

Fig. 2.2 HMM-based phone model.

transition from state $\mathbf{s}_i$ to state $\mathbf{s}_j$ is given by the transition probability $\{a_{ij}\}$. On entering a state, a feature vector is generated using the distribution associated with the state being entered, $\{b_j()\}$. This form of process yields the standard conditional independence assumptions for an HMM:

- states are conditionally independent of all other states given the previous state;
- observations are conditionally independent of all other observations given the state that generated it.

For a more detailed discussion of the operation of an HMM see [141].

For now, single multivariate Gaussians will be assumed for the output distribution:

$$b_j(\boldsymbol{y}) = \mathcal{N}(\boldsymbol{y}; \boldsymbol{\mu}^{(j)}, \boldsymbol{\Sigma}^{(j)}), \tag{2.7}$$

where $\boldsymbol{\mu}^{(j)}$ is the mean of state $\mathbf{s}_j$ and $\boldsymbol{\Sigma}^{(j)}$ is its covariance. Since the dimensionality of the acoustic vector $\boldsymbol{y}$ is relatively high, the covariances are often constrained to be diagonal. Later in *HMM Structure Refinements*, the benefits of using mixtures of Gaussians will be discussed.

Given the composite HMM $\boldsymbol{Q}$ formed by concatenating all of the constituent base phones $\boldsymbol{q}^{(w_1)}, \ldots, \boldsymbol{q}^{(w_L)}$ then the acoustic likelihood is given by

$$p(\boldsymbol{Y}|\boldsymbol{Q}) = \sum_{\theta} p(\boldsymbol{\theta}, \boldsymbol{Y}|\boldsymbol{Q}), \tag{2.8}$$

where $\boldsymbol{\theta} = \theta_0, \ldots, \theta_{T+1}$ is a state sequence through the composite model and

$$p(\boldsymbol{\theta}, \boldsymbol{Y} | \boldsymbol{Q}) = a_{\theta_0 \theta_1} \prod_{t=1}^{T} b_{\theta_t}(\boldsymbol{y}_t) a_{\theta_t \theta_{t+1}}. \tag{2.9}$$

In this equation, θ_0 and θ_{T+1} correspond to the non-emitting entry and exit states shown in Figure 2.2. These are included to simplify the process of concatenating phone models to make words. For simplicity in what follows, these non-emitting states will be ignored and the focus will be on the state sequence $\theta_1, \ldots, \theta_T$.

The acoustic model parameters $\boldsymbol{\lambda} = [\{a_{ij}\}, \{b_j()\}]$ can be efficiently estimated from a corpus of training utterances using the forward–backward algorithm [14] which is an example of expectation-maximisation (EM) [33]. For each utterance $\boldsymbol{Y}^{(r)}, r = 1, \ldots, R$, of length $T^{(r)}$ the sequence of baseforms, the HMMs that correspond to the word-sequence in the utterance, is found and the corresponding composite HMM constructed. In the first phase of the algorithm, the *E-step*, the *forward probability* $\alpha_t^{(rj)} = p(\boldsymbol{Y}_{1:t}^{(r)}, \theta_t = \mathbf{s}_j; \boldsymbol{\lambda})$ and the *backward probability* $\beta_t^{(ri)} = p(\boldsymbol{Y}_{t+1:T^{(r)}}^{(r)} | \theta_t = \mathbf{s}_i; \boldsymbol{\lambda})$ are calculated via the following recursions

$$\alpha_t^{(rj)} = \left[\sum_i \alpha_{t-1}^{(ri)} a_{ij} \right] b_j\left(\boldsymbol{y}_t^{(r)}\right) \tag{2.10}$$

$$\beta_t^{(ri)} = \left[\sum_j a_{ij} b_j(\boldsymbol{y}_{t+1}^{(r)}) \beta_{t+1}^{(rj)} \right], \tag{2.11}$$

where i and j are summed over all states. When performing these recursions under-flow can occur for long speech segments, hence in practice the log-probabilities are stored and *log arithmetic* is used to avoid this problem [89].[5]

Given the forward and backward probabilities, the probability of the model occupying state $\mathbf{s}_j$ at time t for any given utterance r is just

$$\gamma_t^{(rj)} = P(\theta_t = \mathbf{s}_j | \boldsymbol{Y}^{(r)}; \boldsymbol{\lambda}) = \frac{1}{P^{(r)}} \alpha_t^{(rj)} \beta_t^{(rj)}, \tag{2.12}$$

---

[5] This is the technique used in HTK [189].

where $P^{(r)} = p(\boldsymbol{Y}^{(r)}; \boldsymbol{\lambda})$. These state *occupation probabilities*, also called *occupation counts*, represent a soft alignment of the model states to the data and it is straightforward to show that the new set of Gaussian parameters defined by [83]

$$\hat{\boldsymbol{\mu}}^{(j)} = \frac{\sum_{r=1}^{R} \sum_{t=1}^{T^{(r)}} \gamma_t^{(rj)} \boldsymbol{y}_t^{(r)}}{\sum_{r=1}^{R} \sum_{t=1}^{T^{(r)}} \gamma_t^{(rj)}} \tag{2.13}$$

$$\hat{\boldsymbol{\Sigma}}^{(j)} = \frac{\sum_{r=1}^{R} \sum_{t=1}^{T^{(r)}} \gamma_t^{(rj)} (\boldsymbol{y}_t^{(r)} - \hat{\boldsymbol{\mu}}^{(j)})(\boldsymbol{y}_t^{(r)} - \hat{\boldsymbol{\mu}}^{(j)})^{\mathsf{T}}}{\sum_{r=1}^{R} \sum_{t=1}^{T^{(r)}} \gamma_t^{(rj)}} \tag{2.14}$$

maximise the likelihood of the data given these alignments. A similar re-estimation equation can be derived for the transition probabilities

$$\hat{a}_{ij} = \frac{\sum_{r=1}^{R} \frac{1}{P^{(r)}} \sum_{t=1}^{T^{(r)}} \alpha_t^{(ri)} a_{ij} b_j(\boldsymbol{y}_{t+1}^{(r)}) \beta_{t+1}^{(rj)} \boldsymbol{y}_t^{(r)}}{\sum_{r=1}^{R} \sum_{t=1}^{T^{(r)}} \gamma_t^{(ri)}}. \tag{2.15}$$

This is the second or *M-step* of the algorithm. Starting from some initial estimate of the parameters, $\boldsymbol{\lambda}^{(0)}$, successive iterations of the EM algorithm yield parameter sets $\boldsymbol{\lambda}^{(1)}, \boldsymbol{\lambda}^{(2)}, \ldots$ which are guaranteed to improve the likelihood up to some local maximum. A common choice for the initial parameter set $\boldsymbol{\lambda}^{(0)}$ is to assign the global mean and covariance of the data to the Gaussian output distributions and to set all transition probabilities to be equal. This gives a so-called *flat start* model.

This approach to acoustic modelling is often referred to as the *beads-on-a-string* model, so-called because all speech utterances are represented by concatenating a sequence of phone models together. The major problem with this is that decomposing each vocabulary word into a sequence of context-independent base phones fails to capture the very large degree of context-dependent variation that exists in real speech. For example, the base form pronunciations for "mood" and "cool" would use the same vowel for "oo," yet in practice the realisations of "oo" in the two contexts are very different due to the influence of the preceding and following consonant. Context independent phone models are referred to as *monophones*.

A simple way to mitigate this problem is to use a unique phone model for every possible pair of left and right neighbours. The resulting models are called *triphones* and if there are N base phones, there are N^3 potential triphones. To avoid the resulting data sparsity problems, the complete set of *logical* triphones L can be mapped to a reduced set of physical models P by clustering and tying together the parameters in each cluster. This mapping process is illustrated in Figure 2.3 and the parameter tying is illustrated in Figure 2.4 where the notation $x - q + y$ denotes the triphone corresponding to phone q spoken in the context of a preceding phone x and a following phone y. Each base

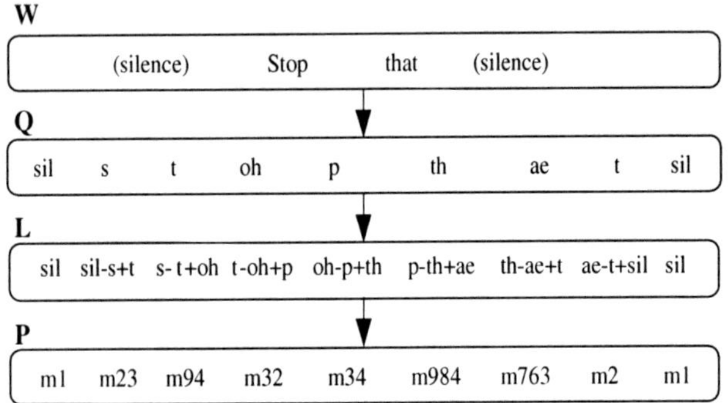

Fig. 2.3 Context dependent phone modelling.

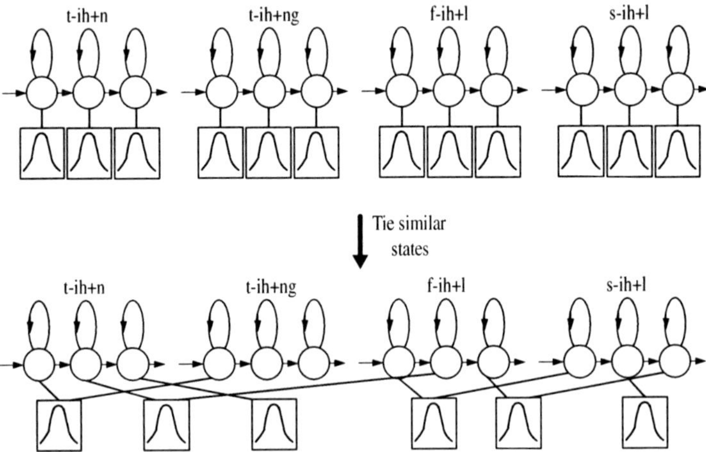

Fig. 2.4 Formation of tied-state phone models.

phone pronunciation q is derived by simple look-up from the pronunciation dictionary, these are then mapped to logical phones according to the context, finally the logical phones are mapped to physical models. Notice that the context-dependence spreads across word boundaries and this is essential for capturing many important phonological processes. For example, the /p/ in "stop that" has its burst suppressed by the following consonant.

The clustering of logical to physical models typically operates at the state-level rather than the model level since it is simpler and it allows a larger set of physical models to be robustly estimated. The choice of which states to tie is commonly made using decision trees [190]. Each state position[6] of each phone q has a binary tree associated with it. Each node of the tree carries a question regarding the context. To cluster state i of phone q, all states i of all of the logical models derived from q are collected into a single pool at the root node of the tree. Depending on the answer to the question at each node, the pool of states is successively split until all states have trickled down to leaf nodes. All states in each leaf node are then tied to form a physical model. The questions at each node are selected from a predetermined set to maximise the likelihood of the training data given the final set of state-tyings. If the state output distributions are single component Gaussians and the state occupation counts are known, then the increase in likelihood achieved by splitting the Gaussians at any node can be calculated simply from the counts and model parameters without reference to the training data. Thus, the decision trees can be grown very efficiently using a greedy iterative node splitting algorithm. Figure 2.5 illustrates this tree-based clustering. In the figure, the logical phones s-aw+n and t-aw+n will both be assigned to leaf node 3 and hence they will share the same central state of the representative physical model.[7]

The partitioning of states using phonetically driven decision trees has several advantages. In particular, logical models which are required but were not seen at all in the training data can be easily synthesised. One disadvantage is that the partitioning can be rather coarse. This

---

[6] Usually each phone model has three states.

[7] The total number of tied-states in a large vocabulary speaker independent system typically ranges between 1000 and 10,000 states.

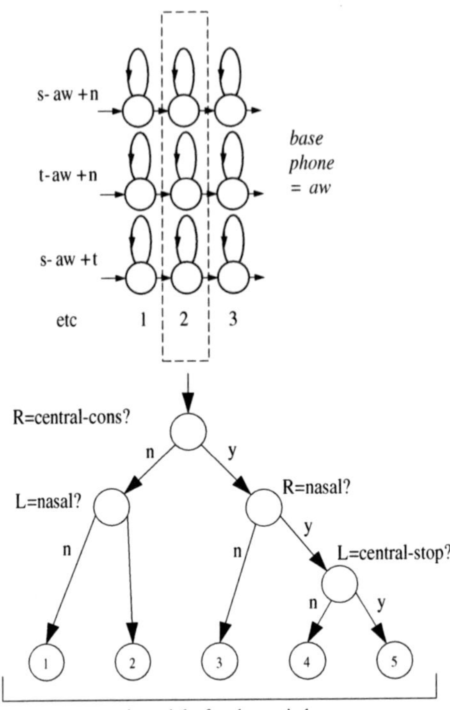

Fig. 2.5 Decision tree clustering.

problem can be reduced using so-called *soft-tying* [109]. In this scheme, a post-processing stage groups each state with its one or two nearest neighbours and pools all of their Gaussians. Thus, the single Gaussian models are converted to mixture Gaussian models whilst holding the total number of Gaussians in the system constant.

To summarise, the core acoustic models of a modern speech recogniser typically consist of a set of tied three-state HMMs with Gaussian output distributions. This core is commonly built in the following steps [189, Ch. 3]:

(1) A flat-start monophone set is created in which each base phone is a monophone single-Gaussian HMM with means and covariances equal to the mean and covariance of the training data.

(2) The parameters of the Gaussian monophones are re-estimated using 3 or 4 iterations of EM.

(3) Each single Gaussian monophone q is cloned once for each distinct triphone $x - q + y$ that appears in the training data.

(4) The resulting set of *training-data* triphones is again re-estimated using EM and the state occupation counts of the last iteration are saved.

(5) A decision tree is created for each state in each base phone, the training-data triphones are mapped into a smaller set of tied-state triphones and iteratively re-estimated using EM.

The final result is the required tied-state context-dependent acoustic model set.

2.3 *N*-gram Language Models

The prior probability of a word sequence $\boldsymbol{w} = w_1, \ldots, w_K$ required in (2.2) is given by

$$P(\boldsymbol{w}) = \prod_{k=1}^{K} P(w_k | w_{k-1}, \ldots, w_1). \tag{2.16}$$

For large vocabulary recognition, the conditioning word history in (2.16) is usually truncated to $N - 1$ words to form an *N-gram* language model

$$P(\boldsymbol{w}) = \prod_{k=1}^{K} P(w_k | w_{k-1}, w_{k-2}, \ldots, w_{k-N+1}), \tag{2.17}$$

where N is typically in the range 2–4. Language models are often assessed in terms of their *perplexity*, H, which is defined as

$$H = -\lim_{K \to \infty} \frac{1}{K} \log_2(P(w_1, \ldots, w_K))$$

$$\approx -\frac{1}{K} \sum_{k=1}^{K} \log_2(P(w_k | w_{k-1}, w_{k-2}, \ldots, w_{k-N+1})),$$

where the approximation is used for N-gram language models with a finite length word sequence.

The N-gram probabilities are estimated from training texts by counting N-gram occurrences to form maximum likelihood (ML) parameter estimates. For example, let $C(w_{k-2}w_{k-1}w_k)$ represent the number of occurrences of the three words $w_{k-2}w_{k-1}w_k$ and similarly for $C(w_{k-2}w_{k-1})$, then

$$P(w_k|w_{k-1}, w_{k-2}) \approx \frac{C(w_{k-2}w_{k-1}w_k)}{C(w_{k-2}w_{k-1})}. \tag{2.18}$$

The major problem with this simple ML estimation scheme is data sparsity. This can be mitigated by a combination of discounting and backing-off. For example, using so-called *Katz smoothing* [85]

$$P(w_k|w_{k-1}, w_{k-2}) = \begin{cases} d\frac{C(w_{k-2}w_{k-1}w_k)}{C(w_{k-2}w_{k-1})} & \text{if } 0 < C \leq C' \\ \frac{C(w_{k-2}w_{k-1}w_k)}{C(w_{k-2}w_{k-1})} & \text{if } C > C' \\ \alpha(w_{k-1}, w_{k-2}) \, P(w_k|w_{k-1}) & \text{otherwise,} \end{cases} \tag{2.19}$$

where C' is a count threshold, C is short-hand for $C(w_{k-2}w_{k-1}w_k)$, d is a discount coefficient and α is a normalisation constant. Thus, when the N-gram count exceeds some threshold, the ML estimate is used. When the count is small the same ML estimate is used but discounted slightly. The discounted probability mass is then distributed to the unseen N-grams which are approximated by a weighted version of the corresponding bigram. This idea can be applied recursively to estimate any sparse N-gram in terms of a set of back-off weights and $(N-1)$-grams. The discounting coefficient is based on the Turing-Good estimate $d = (r+1)n_{r+1}/rn_r$ where n_r is the number of N-grams that occur exactly r times in the training data. There are many variations on this approach [25]. For example, when training data is very sparse, Kneser–Ney smoothing is particularly effective [127].

An alternative approach to robust language model estimation is to use class-based models in which for every word w_k there is a corresponding class c_k [21, 90]. Then,

$$P(\boldsymbol{w}) = \prod_{k=1}^{K} P(w_k|c_k)p(c_k|c_{k-1}, \ldots, c_{k-N+1}). \tag{2.20}$$

As for word based models, the class N-gram probabilities are estimated using ML but since there are far fewer classes (typically a few hundred) data sparsity is much less of an issue. The classes themselves are chosen to optimise the likelihood of the training set assuming a bigram class model. It can be shown that when a word is moved from one class to another, the change in perplexity depends only on the counts of a relatively small number of bigrams. Hence, an iterative algorithm can be implemented which repeatedly scans through the vocabulary, testing each word to see if moving it to some other class would increase the likelihood [115].

In practice it is found that for reasonably sized training sets,[8] an effective language model for large vocabulary applications consists of a smoothed word-based 3 or 4-gram interpolated with a class-based trigram.

2.4   Decoding and Lattice Generation

As noted in the introduction to this section, the most likely word sequence $\hat{w}$ given a sequence of feature vectors $Y_{1:T}$ is found by searching all possible state sequences arising from all possible word sequences for the sequence which was most likely to have generated the observed data $Y_{1:T}$. An efficient way to solve this problem is to use dynamic programming. Let $\phi_t^{(j)} = \max_\theta \{p(Y_{1:t}, \theta_t = s_j; \lambda)\}$, i.e., the maximum probability of observing the partial sequence $Y_{1:t}$ and then being in state s_j at time t given the model parameters λ. This probability can be efficiently computed using the Viterbi algorithm [177]

$$\phi_t^{(j)} = \max_i \left\{ \phi_{t-1}^{(i)} a_{ij} \right\} b_j(y_t). \tag{2.21}$$

It is initialised by setting $\phi_0^{(j)}$ to 1 for the initial, non-emitting, entry state and 0 for all other states. The probability of the most likely word sequence is then given by $\max_j\{\phi_T^{(j)}\}$ and if every maximisation decision is recorded, a traceback will yield the required best matching state/word sequence.

In practice, a direct implementation of the Viterbi algorithm becomes unmanageably complex for continuous speech where the topology of the models, the language model constraints and the need to

---

[8] i.e. $\gg 10^7$ words.

bound the computation must all be taken into account. N-gram language models and cross-word triphone contexts are particularly problematic since they greatly expand the search space. To deal with this, a number of different architectural approaches have evolved. For Viterbi decoding, the search space can either be constrained by maintaining multiple hypotheses in parallel [173, 191, 192] or it can be expanded dynamically as the search progresses [7, 69, 130, 132]. Alternatively, a completely different approach can be taken where the breadth-first approach of the Viterbi algorithm is replaced by a depth-first search. This gives rise to a class of recognisers called *stack decoders*. These can be very efficient, however, because they must compare hypotheses of different lengths, their run-time search characteristics can be difficult to control [76, 135]. Finally, recent advances in weighted finite-state transducer technology enable all of the required information (acoustic models, pronunciation, language model probabilities, etc.) to be integrated into a single very large but highly optimised network [122]. This approach offers both flexibility and efficiency and is therefore extremely useful for both research and practical applications.

Although decoders are designed primarily to find the solution to (2.21), in practice, it is relatively simple to generate not just the most likely hypothesis but the N-best set of hypotheses. N is usually in the range 100–1000. This is extremely useful since it allows multiple passes over the data without the computational expense of repeatedly solving (2.21) from scratch. A compact and efficient structure for storing these hypotheses is the *word lattice* [144, 167, 187].

A word lattice consists of a set of nodes representing points in time and a set of spanning arcs representing word hypotheses. An example is shown in Figure 2.6 part (a). In addition to the word IDs shown in the figure, each arc can also carry score information such as the acoustic and language model scores.

Lattices are extremely flexible. For example, they can be rescored by using them as an input recognition network and they can be expanded to allow rescoring by a higher order language model. They can also be compacted into a very efficient representation called a *confusion network* [42, 114]. This is illustrated in Figure 2.6 part (b) where the "-" arc labels indicate null transitions. In a confusion network, the nodes no

(a) Word Lattice

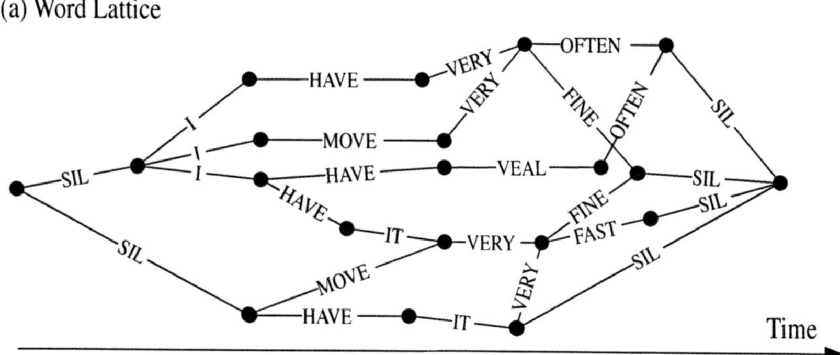

(b) Confusion Network

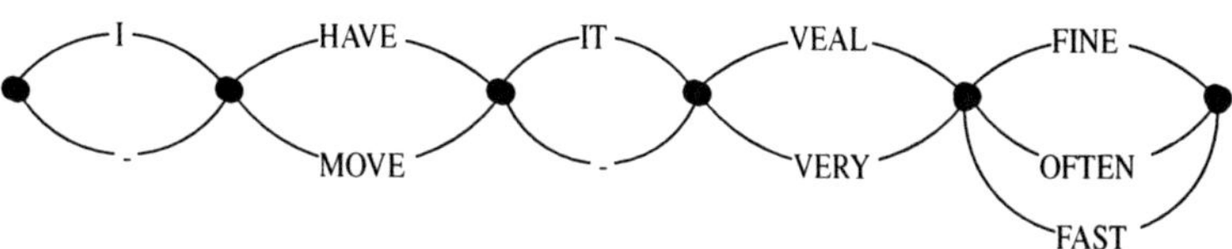

Fig. 2.6 Example lattice and confusion network.

longer correspond to discrete points in time, instead they simply enforce word sequence constraints. Thus, parallel arcs in the confusion network do not necessarily correspond to the same acoustic segment. However, it is assumed that most of the time the overlap is sufficient to enable parallel arcs to be regarded as competing hypotheses. A confusion network has the property that for every path through the original lattice, there exists a corresponding path through the confusion network. Each arc in the confusion network carries the posterior probability of the corresponding word w. This is computed by finding the *link probability* of w in the lattice using a forward–backward procedure, summing over all occurrences of w and then normalising so that all competing word arcs in the confusion network sum to one. Confusion networks can be used for minimum word-error decoding [165] (an example of minimum Bayes' risk (MBR) decoding [22]), to provide confidence scores and for merging the outputs of different decoders [41, 43, 63, 72] (see *Multi-Pass Recognition Architectures*).

3

HMM Structure Refinements

In the previous section, basic acoustic HMMs and their use in ASR systems have been explained. Although these simple HMMs may be adequate for small vocabulary and similar limited complexity tasks, they do not perform well when used for more complex, and larger vocabulary tasks such as broadcast news transcription and dictation. This section describes some of the extensions that have been used to improve the performance of ASR systems and allow them to be applied to these more interesting and challenging domains.

The use of dynamic Bayesian networks to describe possible extensions is first introduced. Some of these extensions are then discussed in detail. In particular, the use of Gaussian mixture models, efficient covariance models and feature projection schemes are presented. Finally, the use of HMMs for generating, rather than recognising, speech is briefly discussed.

3.1 Dynamic Bayesian Networks

In *Architecture of an HMM-Based Recogniser*, the HMM was described as a generative model which for a typical phone has three emitting

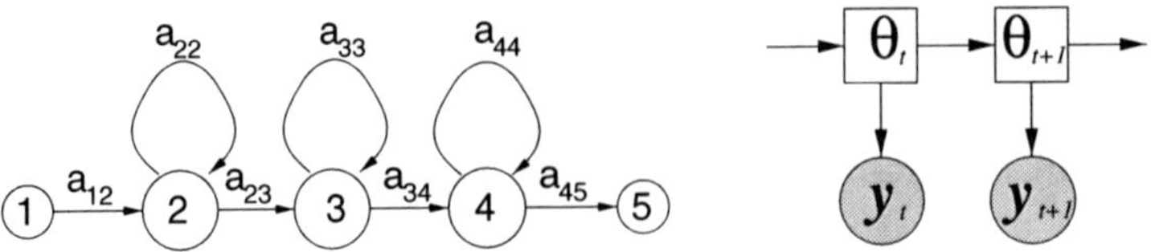

Fig. 3.1 Typical phone HMM topology (left) and dynamic Bayesian network (right).

states, as shown again in Figure 3.1. Also shown in Figure 3.1 is an alternative, complementary, graphical representation called a *dynamic Bayesian network* (*DBN*) which emphasises the conditional dependencies of the model [17, 200] and which is particularly useful for describing a variety of extensions to the basic HMM structure. In the DBN notation used here, squares denote discrete variables; circles continuous variables; shading indicates an observed variable; and no shading an unobserved variable. The lack of an arc between variables shows conditional independence. Thus Figure 3.1 shows that the observations generated by an HMM are conditionally independent given the unobserved, hidden, state that generated it.

One of the desirable attributes of using DBNs is that it is simple to show extensions in terms of how they modify the conditional independence assumptions of the model. There are two approaches, which may be combined, to extend the HMM structure in Figure 3.1: adding additional un-observed variables; and adding additional dependency arcs between variables.

An example of adding additional arcs, in this case between observations is shown in Figure 3.2. Here the observation distribution is

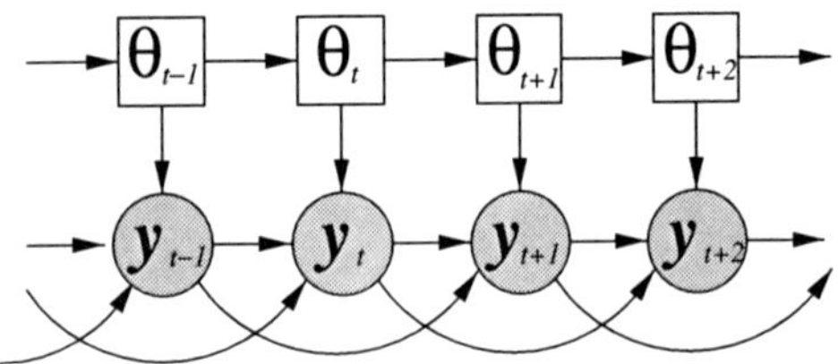

Fig. 3.2 Dynamic Bayesian networks for buried Markov models and HMMs with vector linear predictors.

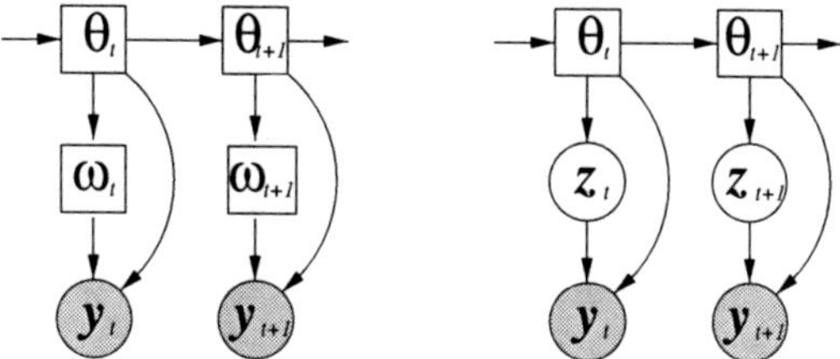

Fig. 3.3 Dynamic Bayesian networks for Gaussian mixture models (left) and factor-analysed models (right).

dependent on the previous two observations in addition to the state that generated it. This DBN describes HMMs with explicit temporal correlation modelling [181], vector predictors [184], and buried Markov models [16]. Although an interesting direction for refining an HMM, this approach has not yet been adopted in mainstream state-of-the-art systems.

Instead of adding arcs, additional unobserved, latent or hidden, variables may be added. To compute probabilities, these hidden variables must then be marginalised out in the same fashion as the unobserved state sequence for HMMs. For continuous variables this requires a continuous integral over all values of the hidden variable and for discrete variables a summation over all values. Two possible forms of latent variable are shown in Figure 3.3. In both cases the temporal dependencies of the model, where the discrete states are conditionally independent of all other states given the previous state, are unaltered allowing the use of standard Viterbi decoding routines. In the first case, the observations are dependent on a discrete latent variable. This is the DBN for an HMM with Gaussian mixture model (GMM) state-output distributions. In the second case, the observations are dependent on a continuous latent variable. This is the DBN for an HMM with factor-analysed covariance matrices. Both of these are described in more detail below.

3.2 Gaussian Mixture Models

One of the most commonly used extensions to standard HMMs is to model the state-output distribution as a mixture model. In *Architecture of an HMM-Based Recogniser*, a single Gaussian distribution was used to model the state–output distribution. This model therefore assumed that the observed feature vectors are symmetric and unimodal. In practice this is seldom the case. For example, speaker, accent and gender differences tend to create multiple modes in the data. To address this problem, the single Gaussian state-output distribution may be replaced by a mixture of Gaussians which is a highly flexible distribution able to model, for example, asymmetric and multi-modal distributed data.

As described in the previous section, mixture models may be viewed as adding an additional discrete latent variable to the system. The likelihood for state $\mathbf{s}_j$ is now obtained by summing over all the component likelihoods weighted by their prior

$$b_j(\boldsymbol{y}) = \sum_{m=1}^{M} c_{jm}\mathcal{N}(\boldsymbol{y};\boldsymbol{\mu}^{(jm)},\boldsymbol{\Sigma}^{(jm)}), \qquad (3.1)$$

where c_{jm} is the prior probability for component m of state $\mathbf{s}_j$. These priors satisfy the standard constraints for a valid probability mass function (PMF)

$$\sum_{m=1}^{M} c_{jm} = 1, \quad c_{jm} \geq 0. \qquad (3.2)$$

This is an example of the general technique of mixture modelling. Each of the M components of the mixture model is a Gaussian probability density function (PDF).

Since an additional latent variable has been added to the acoustic model, the form of the EM algorithm previously described needs to be modified [83]. Rather than considering the complete data-set in terms of state-observation pairings, state/component-observations are used. The estimation of the model parameters then follows the same form as for the single component case. For the mean of Gaussian component m of state $\mathbf{s}_j$, $\mathbf{s}_{jm}$,

$$\hat{\boldsymbol{\mu}}^{(jm)} = \frac{\sum_{r=1}^{R}\sum_{t=1}^{T^{(r)}} \gamma_t^{(rjm)}\boldsymbol{y}_t^{(r)}}{\sum_{r=1}^{R}\sum_{t=1}^{T^{(r)}} \gamma_t^{(rjm)}}, \qquad (3.3)$$

where $\gamma_t^{(rjm)} = P(\theta_t = \mathbf{s}_{jm}|\boldsymbol{Y}^{(r)};\boldsymbol{\lambda})$ is the probability that component m of state $\mathbf{s}_j$ generated the observation at time t of sequence $\boldsymbol{Y}^{(r)}$.

R is the number of training sequences. The component priors are estimated in a similar fashion to the transition probabilities.

When using GMMs to model the state-output distribution, variance flooring is often applied.[1] This prevents the variances of the system becoming too small, for example when a component models a very small number of tightly packed observations. This improves generalisation.

Using GMMs increases the computational overhead since when using log-arithmetic, a series of log-additions are required to compute the GMM likelihood. To improve efficiency, only components which provide a "reasonable" contribution to the total likelihood are included. To further decrease the computational load the GMM likelihood can be approximated simply by the maximum over all the components (weighted by the priors).

In addition, it is necessary to determine the number of components per state in the system. A number of approaches, including discriminative schemes, have been adopted for this [27, 106]. The simplest is to use the same number of components for all states in the system and use held-out data to determine the optimal numbers. A popular alternative approach is based on the Bayesian information criterion (BIC) [27]. Another approach is to make the number of components assigned to the state a function of the number of observations assigned to that state [56].

3.3 Feature Projections

In *Architecture for an HMM-Based Recogniser*, dynamic first and second differential parameters, the so-called *delta* and *delta–delta* parameters, were added to the static feature parameters to overcome the limitations of the conditional independence assumption associated with HMMs. Furthermore, the DCT was assumed to approximately decorrelate the feature vector to improve the diagonal covariance approximation and reduce the dimensionality of the feature vector.

It is also possible to use data-driven approaches to decorrelate and reduce the dimensionality of the features. The standard approach is to

---

[1] See HTK for a specific implementation of variance flooring [189].

use linear transformations, that is

$$\boldsymbol{y}_t = \mathbf{A}_{[p]}\tilde{\boldsymbol{y}}_t, \tag{3.4}$$

where $\mathbf{A}_{[p]}$ is a $p \times d$ linear-transform, d is the dimension of the source feature vector $\tilde{\boldsymbol{y}}_t$, and p is the size of the transformed feature vector $\boldsymbol{y}_t$.

The detailed form of a feature vector transformation such as this depends on a number of choices. Firstly, the construction of $\tilde{\boldsymbol{y}}_t$ must be decided and, when required, the class labels that are to be used, for example phone, state or Gaussian component labels. Secondly, the dimensionality p of the projected data must be determined. Lastly, the criterion used to estimate $\mathbf{A}_{[p]}$ must be specified.

An important issue when estimating a projection is whether the class labels of the observations are to be used, or not, i.e., whether the projection should be estimated in a *supervised* or *unsupervised* fashion. In general supervised approaches yield better projections since it is then possible to estimate a transform to improve the discrimination between classes. In contrast, in unsupervised approaches only general attributes of the observations, such as the variances, may be used. However, supervised schemes are usually computationally more expensive than unsupervised schemes. Statistics, such as the means and covariance matrices are needed for each of the class labels, whereas unsupervised schemes effectively only ever have one class.

The features to be projected are normally based on the standard MFCC, or PLP, feature-vectors. However, the treatment of the delta and delta–delta parameters can vary. One approach is to splice neighbouring static vectors together to form a composite vector of typically 9 frames [10]

$$\tilde{\boldsymbol{y}}_t = \begin{bmatrix} \boldsymbol{y}_{t-4}^{\mathsf{s}\mathsf{T}} & \cdots & \boldsymbol{y}_t^{\mathsf{s}\mathsf{T}} & \cdots & \boldsymbol{y}_{t+4}^{\mathsf{s}\mathsf{T}} \end{bmatrix}^{\mathsf{T}}. \tag{3.5}$$

Another approach is to expand the static, delta and delta–delta parameters with third-order dynamic parameters (the difference of delta–deltas) [116]. Both have been used for large vocabulary speech recognition systems. The dimensionality of the projected feature vector is often determined empirically since there is only a single parameter to tune. In [68] a comparison of a number of possible class definitions, phone, state or component, were compared for supervised projection schemes. The choice of class is important as the transform tries to ensure that each of these classes is as separable as possible from all others. The study showed that performance using state and component level classes was similar, and both were better than higher level labels such as phones or words. In practice, the vast majority of systems use component level classes since, as discussed later, this improves the assumption of diagonal covariance matrices.

The simplest form of criterion for estimating the transform is principal component analysis (PCA). This is an unsupervised projection scheme, so no use is made of class labels. The estimation of the PCA transform can be found by finding the p rows of the orthonormal matrix $\mathbf{A}$ that maximises

$$\mathcal{F}_{\text{pca}}(\boldsymbol{\lambda}) = \log\left(|\mathbf{A}_{[p]}\tilde{\boldsymbol{\Sigma}}_{\text{g}}\mathbf{A}_{[p]}^{\mathsf{T}}|\right), \tag{3.6}$$

where $\tilde{\boldsymbol{\Sigma}}_{\text{g}}$ is the total, or global, covariance matrix of the original data, $\tilde{\boldsymbol{y}}_t$. This selects the orthogonal projections of the data that maximises the total variance in the projected subspace. Simply selecting subspaces that yield large variances does not necessarily yield subspaces that discriminate between the classes. To address this problem, supervised approaches such as the linear discriminant analysis (LDA) criterion [46] can be used. In LDA the objective is to increase the ratio of the between class variance to the average within class variance for each dimension. This criterion may be expressed as

$$\mathcal{F}_{\text{lda}}(\boldsymbol{\lambda}) = \log\left(\frac{|\mathbf{A}_{[p]}\tilde{\boldsymbol{\Sigma}}_{\text{b}}\mathbf{A}_{[p]}^{\mathsf{T}}|}{|\mathbf{A}_{[p]}\tilde{\boldsymbol{\Sigma}}_{\text{w}}\mathbf{A}_{[p]}^{\mathsf{T}}|}\right), \tag{3.7}$$

where $\tilde{\boldsymbol{\Sigma}}_{\text{b}}$ is the between-class covariance matrix and $\tilde{\boldsymbol{\Sigma}}_{\text{w}}$ the average within-class covariance matrix where each distinct Gaussian component is usually assumed to be a separate class. This criterion yields an orthonormal transform such that the average within-class covariance matrix is diagonalised, which should improve the diagonal covariance matrix assumption.

The LDA criterion may be further refined by using the actual class covariance matrices rather than using the averages. One such form is heteroscedastic discriminant analysis (HDA) [149] where the following criterion is maximised

$$\mathcal{F}_{\mathsf{hda}}(\boldsymbol{\lambda}) = \sum_m \gamma^{(m)} \log \left(\frac{|\mathbf{A}_{[p]}\tilde{\boldsymbol{\Sigma}}_{\mathsf{b}}\mathbf{A}_{[p]}^{\mathsf{T}}|}{|\mathbf{A}_{[p]}\tilde{\boldsymbol{\Sigma}}^{(m)}\mathbf{A}_{[p]}^{\mathsf{T}}|} \right), \tag{3.8}$$

where $\gamma^{(m)}$ is the total posterior occupation probability for component m and $\tilde{\boldsymbol{\Sigma}}^{(m)}$ is its covariance matrix in the original space given by $\tilde{\boldsymbol{y}}_t$. In this transformation, the matrix $\mathbf{A}$ is not constrained to be orthonormal, and nor does this criterion help with decorrelating the data associated with each Gaussian component. It is thus often used in conjunction with a global semi-tied transform [51] (also known as a maximum likelihood linear transform (MLLT) [65]) described in the next section. An alternative extension to LDA is heteroscedastic LDA (HLDA) [92]. This modifies the LDA criterion in (3.7) in a similar fashion to HDA, but now a transform for the complete feature-space is estimated, rather than for just the dimensions to be retained. The parameters of the HLDA transform can be found in an ML fashion, as if they were model parameters as discussed in *Architecture for an HMM-Based Recogniser*. An important extension of this transform is to ensure that the distributions for all dimensions to be removed are constrained to be the same. This is achieved by tying the parameters associated with these dimensions to ensure that they are identical. These dimensions will thus yield no discriminatory information and so need not be retained for recognition. The HLDA criterion can then be expressed as

$$\mathcal{F}_{\mathsf{hlda}}(\boldsymbol{\lambda}) = \sum_m \gamma^{(m)} \log \left(\frac{|\mathbf{A}|^2}{\mathrm{diag}\left(|\mathbf{A}_{[d-p]}\tilde{\boldsymbol{\Sigma}}_{\mathsf{g}}\mathbf{A}_{[d-p]}^{\mathsf{T}}|\right) \mathrm{diag}\left(|\mathbf{A}_{[p]}\tilde{\boldsymbol{\Sigma}}^{(m)}\mathbf{A}_{[p]}^{\mathsf{T}}|\right)} \right), \tag{3.9}$$

where

$$\mathbf{A} = \begin{bmatrix} \mathbf{A}_{[p]} \\ \mathbf{A}_{[d-p]} \end{bmatrix}. \tag{3.10}$$

There are two important differences between HLDA and HDA. Firstly, HLDA yields the best projection whilst simultaneously generating the best transform for improving the diagonal covariance matrix approximation. In contrast, for HDA a separate decorrelating transform must be added. Furthermore, HLDA yields a model for the complete feature-space, whereas HDA only models the useful (non-projected) dimensions. This means that multiple subspace projections can be used with HLDA, but not with HDA [53].

Though schemes like HLDA out-perform approaches like LDA they are more computationally expensive and require more memory. Full-covariance matrix statistics for each component are required to estimate an HLDA transform, whereas only the average within and between class covariance matrices are required for LDA. This makes HLDA projections from large dimensional features spaces with large numbers of components impractical. One compromise that is sometimes used is to follow an LDA projection by a decorrelating global semi-tied transform [163] described in the next section.

3.4 Covariance Modelling

One of the motivations for using the DCT transform, and some of the projection schemes discussed in the previous section, is to decorrelate the feature vector so that the diagonal covariance matrix approximation becomes reasonable. This is important as in general the use of full covariance Gaussians in large vocabulary systems would be impractical due to the sheer size of the model set.[2] Even with small systems, training data limitations often preclude the use of full covariances. Furthermore, the computational cost of using full covariance matrices is $\mathcal{O}(d^2)$ compared to $\mathcal{O}(d)$ for the diagonal case where d is the dimensionality of the feature vector. To address both of these problems, structured covariance and precision (inverse covariance) matrix representations have been developed. These allow covariance modelling to be improved with very little overhead in terms of memory and computational load.

3.4.1 Structured Covariance Matrices

A standard form of structured covariance matrix arises from factor analysis and the DBN for this was shown in Figure 3.3. Here the covariance matrix for each Gaussian component m is represented by

---

[2] Although given enough data, and some smoothing, it can be done [163].

(the dependency on the state has been dropped for clarity)

$$\mathbf{\Sigma}^{(m)} = \mathbf{A}_{[p]}^{(m)\mathsf{T}} \mathbf{A}_{[p]}^{(m)} + \mathbf{\Sigma}_{\text{diag}}^{(m)}, \tag{3.11}$$

where $\mathbf{A}_{[p]}^{(m)}$ is the component-specific, $p \times d$, *loading matrix* and $\mathbf{\Sigma}_{\text{diag}}^{(m)}$ is a component specific diagonal covariance matrix. The above expression is based on factor analysis which allows each of the Gaussian components to be estimated separately using EM [151]. Factor-analysed HMMs [146] generalise this to support both tying over multiple components of the loading matrices and using GMMs to model the latent variable-space of z in Figure 3.3.

Although structured covariance matrices of this form reduce the number of parameters needed to represent each covariance matrix, the computation of likelihoods depends on the inverse covariance matrix, i.e., the precision matrix and this will still be a full matrix. Thus, factor-analysed HMMs still incur the decoding cost associated with full covariances.[3]

3.4.2   Structured Precision Matrices

A more computationally efficient approach to covariance structuring is to model the inverse covariance matrix and a general form for this is [131, 158]

$$\mathbf{\Sigma}^{(m)-1} = \sum_{i=1}^{B} \nu_i^{(m)} \mathbf{S}_i, \tag{3.12}$$

where $\boldsymbol{\nu}^{(m)}$ is a Gaussian component specific weight vector that specifies the contribution from each of the B global positive semi-definite matrices, $\mathbf{S}_i$. One of the desirable properties of this form of model is that it is computationally efficient during decoding, since

$$\log\left(\mathcal{N}(\boldsymbol{y}; \boldsymbol{\mu}^{(m)}, \mathbf{\Sigma}^{(m)})\right)$$

$$= -\frac{1}{2}\log((2\pi)^d |\mathbf{\Sigma}^{(m)}|) - \frac{1}{2}\sum_{i=1}^{B} \nu_i^{(m)} (\boldsymbol{y} - \boldsymbol{\mu}^{(m)})^\mathsf{T} \mathbf{S}_i (\boldsymbol{y} - \boldsymbol{\mu}^{(m)})$$

---

[3] Depending on p, the matrix inversion lemmas, also known as the Sherman-Morrisson-Woodbury formula, can be used to make this more efficient.

$$= -\frac{1}{2} \log((2\pi)^d |\mathbf{\Sigma}^{(m)}|)$$

$$-\frac{1}{2} \sum_{i=1}^{B} \nu_i^{(m)} \left(\mathbf{y}^\mathsf{T} \mathbf{S}_i \mathbf{y} - 2\boldsymbol{\mu}^{(m)\mathsf{T}} \mathbf{S}_i \mathbf{y} + \boldsymbol{\mu}^{(m)\mathsf{T}} \mathbf{S}_i \boldsymbol{\mu}^{(m)} \right). \quad (3.13)$$

Thus rather than having the computational cost of a full covariance matrix, it is possible to cache terms such as $\mathbf{S}_i \mathbf{y}$ and $\mathbf{y}^\mathsf{T} \mathbf{S}_i \mathbf{y}$ which do not depend on the Gaussian component.

One of the simplest and most effective structured covariance representations is the semi-tied covariance matrix (STC) [51]. This is a specific form of precision matrix model, where the number of bases is equal to the number of dimensions ($B = d$) and the bases are symmetric and have rank 1 (thus can be expressed as $\mathbf{S}_i = \mathbf{a}_{[i]}^\mathsf{T} \mathbf{a}_{[i]}$, $\mathbf{a}_{[i]}$ is the ith *row* of $\mathbf{A}$). In this case the component likelihoods can be computed by

$$\mathcal{N}\left(\mathbf{y}; \boldsymbol{\mu}^{(m)}, \mathbf{\Sigma}^{(m)} \right) = |\mathbf{A}| \mathcal{N}\left(\mathbf{A}\mathbf{y}; \mathbf{A}\boldsymbol{\mu}^{(m)}, \mathbf{\Sigma}_{\text{diag}}^{(m)} \right) \quad (3.14)$$

where $\mathbf{\Sigma}_{\text{diag}}^{(m)-1}$ is a diagonal matrix formed from $\boldsymbol{\nu}^{(m)}$ and

$$\mathbf{\Sigma}^{(m)-1} = \sum_{i=1}^{d} \nu_i^{(m)} \mathbf{a}_{[i]}^\mathsf{T} \mathbf{a}_{[i]} = \mathbf{A} \mathbf{\Sigma}_{\text{diag}}^{(m)-1} \mathbf{A}^\mathsf{T}. \quad (3.15)$$

The matrix $\mathbf{A}$ is sometimes referred to as the semi-tied transform. One of the reasons that STC systems are simple to use is that the weight for each dimension is simply the inverse variance for that dimension. The training procedure for STC systems is (after accumulating the standard EM full-covariance matrix statistics):

(1) initialise the transform $\mathbf{A}^{(0)} = \mathbf{I}$; set $\mathbf{\Sigma}_{\text{diag}}^{(m0)}$ equal to the current model covariance matrices; and set $k = 0$;
(2) estimate $\mathbf{A}^{(k+1)}$ given $\mathbf{\Sigma}_{\text{diag}}^{(mk)}$;
(3) set $\mathbf{\Sigma}_{\text{diag}}^{(m(k+1))}$ to the component variance for each dimension using $\mathbf{A}^{(k+1)}$;
(4) goto (2) unless converged or maximum number of iterations reached.

During recognition, decoding is very efficient since $\mathbf{A}\boldsymbol{\mu}^{(m)}$ is stored for each component, and the transformed features $\mathbf{A}\mathbf{y}_t$ are cached for each time instance. There is thus almost no increase in decoding time compared to standard diagonal covariance matrix systems.

STC systems can be made more powerful by using multiple semi-tied transforms [51]. In addition to STC, other types of structured covariance modelling include sub-space constrained precision and means (SPAM) [8], mixtures of inverse covariances [175], and extended maximum likelihood linear transforms (EMLLT) [131].

3.5   HMM Duration Modelling

The probability $d_j(t)$ of remaining in state s_j for t consecutive observations in a standard HMM is given by

$$d_j(t) = a_{jj}^{t-1}(1 - a_{jj}), \qquad (3.16)$$

where a_{jj} is the self-transition probability of state s_j. Thus, the standard HMM models the probability of state occupancy as decreasing exponentially with time and clearly this is a poor model of duration.

Given this weakness of the standard HMM, an obvious refinement is to introduce an explicit duration model such that the self-transition loops in Figure 2.2 are replaced by an explicit probability distribution $d_j(t)$. Suitable choices for $d_j(t)$ are the Poisson distribution [147] and the Gamma distribution [98].

Whilst the use of these explicit distributions can undoubtably model state segment durations more accurately, they add a very considerable overhead to the computational complexity. This is because the resulting models no longer have the Markov property and this prevents efficient search over possible state alignments. For example, in these so-called *hidden semi-Markov models* (*HSMMs*), the forward probability given in (2.10) becomes

$$\alpha_t^{(rj)} = \sum_{\tau} \sum_{i, i \neq j} \alpha_{t-\tau}^{(ri)} a_{ij} d_j(\tau) \prod_{l=1}^{\tau} b_j \left(y_{t-\tau+l}^{(r)} \right). \qquad (3.17)$$

The need to sum over all possible state durations has increased the complexity by a factor of $O(D^2)$ where D is the maximum allowed state duration, and the backward probability and Viterbi decoding suffer the same increase in complexity.

In practice, explicit duration modelling of this form offers some small improvements to performance in small vocabulary applications, and minimal improvements in large vocabulary applications. A typical experience in the latter is that as the accuracy of the phone-output distributions is increased, the improvements gained from duration modelling become negligible. For this reason and its computational complexity, durational modelling is rarely used in current systems.

3.6 HMMs for Speech Generation

HMMs are generative models and although HMM-based acoustic models were developed primarily for speech recognition, it is relevant to consider how well they can actually generate speech. This is not only of direct interest for synthesis applications where the flexibility and compact representation of HMMs offer considerable benefits [168], but it can also provide further insight into their use in recognition [169].

The key issue of interest here is the extent to which the dynamic delta and delta–delta terms can be used to enforce accurate trajectories for the static parameters. If an HMM is used to directly model the speech features, then given the conditional independence assumptions from Figure 3.1 and a particular state sequence, θ, the trajectory will be piece-wise stationary where the time segment corresponding to each state, simply adopts the mean-value of that state. This would clearly be a poor fit to real speech where the spectral parameters vary much more smoothly.

The *HMM-trajectory model*[4] aims to generate realistic feature trajectories by finding the sequence of d-dimensional static observations which maximises the likelihood of the complete feature vector (i.e. statics + deltas) with respect to the parameters of a standard HMM model. The trajectory of these static parameters will no longer be piecewise stationary since the associated delta parameters also contribute to the likelihood and must therefore be consistent with the HMM parameters.

To see how these trajectories can be computed, consider the case of using a feature vector comprising static parameters and "simple

---

[4] An implementation of this is available as an extension to HTK at `hts.sp.nitech.ac.jp`.

difference" delta parameters. The observation at time t is a function of the static features at times $t - 1$, t and $t + 1$

$$y_t = \begin{bmatrix} y_t^s \\ \Delta y_t^s \end{bmatrix} = \begin{bmatrix} 0 & I & 0 \\ -I & 0 & I \end{bmatrix} \begin{bmatrix} y_{t-1}^s \\ y_t^s \\ y_{t+1}^s \end{bmatrix}, \qquad (3.18)$$

where 0 is a $d \times d$ zero matrix and I is a $d \times d$ identity matrix. y_t^s is the static element of the feature vector at time t and Δy_t^s are the delta features.[5]

Now if a complete sequence of features is considered, the following relationship can be obtained between the $2Td$ features used for the standard HMM and the Td static features

$$Y_{1:T} = A Y^s. \qquad (3.19)$$

To illustrate the form of the $2Td \times Td$ matrix A consider the observation vectors at times, $t - 1$, t and $t + 1$ in the complete sequence. These may be expressed as

$$\begin{bmatrix} \vdots \\ y_{t-1} \\ y_t \\ y_{t+1} \\ \vdots \end{bmatrix} = \begin{bmatrix} \cdots & 0 & I & 0 & 0 & 0 & \cdots \\ \cdots & -I & 0 & I & 0 & 0 & \cdots \\ \cdots & 0 & 0 & I & 0 & 0 & \cdots \\ \cdots & 0 & -I & 0 & I & 0 & \cdots \\ \cdots & 0 & 0 & 0 & I & 0 & \cdots \\ \cdots & 0 & 0 & -I & 0 & I & \cdots \end{bmatrix} \begin{bmatrix} \vdots \\ y_{t-2}^s \\ y_{t-1}^s \\ y_t^s \\ y_{t+1}^s \\ y_{t+2}^s \\ \vdots \end{bmatrix}. \qquad (3.20)$$

The features modelled by the standard HMM are thus a linear transform of the static features. It should therefore be possible to derive the probability distribution of the static feature distribution if the parameters of the standard HMM are known.

The process of finding the distribution of the static features is simplified by hypothesising an appropriate state/component-sequence θ for the observations to be generated. In practical synthesis systems state duration models are often estimated as in the previous section. In this

---

[5] Here a non-standard version of delta parameters is considered to simplify notation. In this case, d is the size of the static parameters, not the complete feature vector.

case the simplest approach is to set the duration of each state equal to its average (note only single component state-output distributions can be used). Given $\boldsymbol{\theta}$, the distribution of the associated static sequences $\boldsymbol{Y}^{\text{s}}$ will be Gaussian distributed. The likelihood of a static sequence can then be expressed in terms of the standard HMM features as

$$\frac{1}{Z}p(\mathbf{A}\boldsymbol{Y}^{\text{s}}|\boldsymbol{\theta};\lambda) = \frac{1}{Z}\mathcal{N}(\boldsymbol{Y}_{1:T};\boldsymbol{\mu}_\theta,\boldsymbol{\Sigma}_\theta) = \mathcal{N}(\boldsymbol{Y}^{\text{s}};\boldsymbol{\mu}_\theta^{\text{s}},\boldsymbol{\Sigma}_\theta^{\text{s}}), \qquad (3.21)$$

where Z is a normalisation term. The following relationships exist between the standard model parameters and the static parameters

$$\boldsymbol{\Sigma}_\theta^{\text{s}-1} = \mathbf{A}^\mathsf{T}\boldsymbol{\Sigma}_\theta^{-1}\mathbf{A} \qquad (3.22)$$

$$\boldsymbol{\Sigma}_\theta^{\text{s}-1}\boldsymbol{\mu}_\theta^{\text{s}} = \mathbf{A}^\mathsf{T}\boldsymbol{\Sigma}_\theta^{-1}\boldsymbol{\mu}_\theta \qquad (3.23)$$

$$\boldsymbol{\mu}_\theta^{\text{s}\mathsf{T}}\boldsymbol{\Sigma}_\theta^{\text{s}-1}\boldsymbol{\mu}_\theta^{\text{s}} = \boldsymbol{\mu}_\theta^\mathsf{T}\boldsymbol{\Sigma}_\theta^{-1}\boldsymbol{\mu}_\theta, \qquad (3.24)$$

where the segment mean, $\boldsymbol{\mu}_\theta$, and variances, $\boldsymbol{\Sigma}_\theta$, (along with the corresponding versions for the static parameters only, $\boldsymbol{\mu}_\theta^{\text{s}}$ and $\boldsymbol{\Sigma}_\theta^{\text{s}}$) are

$$\boldsymbol{\mu}_\theta = \begin{bmatrix} \boldsymbol{\mu}^{(\theta_1)} \\ \vdots \\ \boldsymbol{\mu}^{(\theta_T)} \end{bmatrix}, \quad \boldsymbol{\Sigma}_\theta = \begin{bmatrix} \boldsymbol{\Sigma}^{(\theta_1)} & \cdots & \mathbf{0} \\ \vdots & \ddots & \vdots \\ \mathbf{0} & \cdots & \boldsymbol{\Sigma}^{(\theta_T)} \end{bmatrix}. \qquad (3.25)$$

Though the covariance matrix of the HMM features, $\boldsymbol{\Sigma}_\theta$, has the block-diagonal structure expected from the HMM, the covariance matrix of the static features, $\boldsymbol{\Sigma}_\theta^{\text{s}}$ in (3.22), does not have a block diagonal structure since $\mathbf{A}$ is not block diagonal. Thus, using the delta parameters to obtain the distribution of the static parameters does not imply the same conditional independence assumptions as the standard HMM assumes in modelling the static features.

The maximum likelihood static feature trajectory, $\hat{\boldsymbol{Y}}^{\text{s}}$, can now be found using (3.21). This will simply be the static parameter segment mean, $\boldsymbol{\mu}_\theta^{\text{s}}$, as the distribution is Gaussian. To find $\boldsymbol{\mu}_\theta^{\text{s}}$ the relationships in (3.22) to (3.24) can be rearranged to give

$$\hat{\boldsymbol{Y}}^{\text{s}} = \boldsymbol{\mu}_\theta^{\text{s}} = \left(\mathbf{A}^\mathsf{T}\boldsymbol{\Sigma}_\theta^{-1}\mathbf{A}\right)^{-1}\mathbf{A}^\mathsf{T}\boldsymbol{\Sigma}_\theta^{-1}\boldsymbol{\mu}_\theta. \qquad (3.26)$$

These are all known as $\boldsymbol{\mu}_\theta$ and $\boldsymbol{\Sigma}_\theta$ are based on the standard HMM parameters and the matrix $\mathbf{A}$ is given in (3.25). Note, given the highly structured-nature of $\mathbf{A}$ efficient implementations of the matrix inversion may be found.

The basic HMM synthesis model described has been refined in a couple of ways. If the overall length of the utterance to synthesise is known, then an ML estimate of $\boldsymbol{\theta}$ can be found in a manner similar to Viterbi. Rather than simply using the average state durations for the synthesis process, it is also possible to search for the most likely state/component-sequence. Here

$$\left\{\hat{\boldsymbol{Y}}^{\mathrm{s}}, \hat{\boldsymbol{\theta}}\right\} = \underset{\boldsymbol{Y}^{\mathrm{s}}, \boldsymbol{\theta}}{\arg\max}\left\{\frac{1}{Z}p(\mathbf{A}\boldsymbol{Y}^{\mathrm{s}}|\boldsymbol{\theta};\boldsymbol{\lambda})P(\boldsymbol{\theta};\boldsymbol{\lambda})\right\}. \qquad (3.27)$$

In addition, though the standard HMM model parameters, $\boldsymbol{\lambda}$, can be trained in the usual fashion for example using ML, improved performance can be obtained by training them so that when used to generate sequences of static features they are a "good" model of the training data. ML-based estimates for this can be derived [168]. Just as regular HMMs can be trained discriminatively as described later, HMM-synthesis models can also be trained discriminatively [186]. One of the major issues with both these improvements is that the Viterbi algorithm cannot be directly used with the model to obtain the state-sequence. A frame-delayed version of the Viterbi algorithm can be used [196] to find the state-sequence. However, this is still more expensive than the standard Viterbi algorithm.

HMM trajectory models can also be used for speech recognition [169]. This again makes use of the equalities in (3.21). The recognition output for feature vector sequence $\boldsymbol{Y}_{1:T}$ is now based on

$$\hat{\boldsymbol{w}} = \underset{\boldsymbol{w}}{\arg\max}\left\{\underset{\boldsymbol{\theta}}{\arg\max}\left\{\frac{1}{Z}\mathcal{N}(\boldsymbol{Y}_{1:T};\boldsymbol{\mu}_{\theta},\boldsymbol{\Sigma}_{\theta})P(\boldsymbol{\theta}|\boldsymbol{w};\boldsymbol{\lambda})P(\boldsymbol{w})\right\}\right\}. \quad (3.28)$$

As with synthesis, one of the major issues with this form of decoding is that the Viterbi algorithm cannot be used. In practice N-best list rescoring is often implemented instead. Though an interesting research direction, this form of speech recognition system is not in widespread use.

4

Parameter Estimation

In *Architecture of an HMM-Based Recogniser*, the estimation of the HMM model parameters, $\boldsymbol{\lambda}$, was briefly described based on maximising the likelihood that the models generate the training sequences. Thus, given training data $\boldsymbol{Y}^{(1)}, \ldots, \boldsymbol{Y}^{(R)}$ the maximum likelihood (ML) training criterion may be expressed as

$$\mathcal{F}_{\mathtt{ml}}(\boldsymbol{\lambda}) = \frac{1}{R} \sum_{r=1}^{R} \log(p(\boldsymbol{Y}^{(r)} | \boldsymbol{w}_{\mathtt{ref}}^{(r)}; \boldsymbol{\lambda})), \tag{4.1}$$

where $\boldsymbol{Y}^{(r)}$ is the rth training utterance with transcription $\boldsymbol{w}_{\mathtt{ref}}^{(r)}$. This optimisation is normally performed using EM [33]. However, for ML to be the "best" training criterion, the data and models would need to satisfy a number of requirements, in particular, training data sufficiency and model-correctness [20]. Since in general these requirements are not satisfied when modelling speech data, alternative criteria have been developed.

This section describes the use of discriminative criteria for training HMM model parameters. In these approaches, the goal is not to estimate the parameters that are most likely to generate the training

37

data. Instead, the objective is to modify the model parameters so that hypotheses generated by the recogniser on the training data more closely "match" the correct word-sequences, whilst generalising to unseen test data.

One approach would be to explicitly minimise the classification error on the training data. Here, a strict 1/0 loss function is needed as the measure to optimise. However, the differential of such a loss function would be discontinuous preventing the use of gradient descent based schemes. Hence various approximations described in the next section have been explored. Compared to ML training, the implementation of these training criteria is more complex and they have a tendency to *over-train* on the training data, i.e., they do not generalise well to unseen data. These issues are also discussed.

One of the major costs incurred in building speech recognition systems lies in obtaining sufficient acoustic data with accurate transcriptions. Techniques which allow systems to be trained with rough, approximate transcriptions, or even no transcriptions at all are therefore of growing interest and work in this area is briefly summarised at the end of this section.

4.1 Discriminative Training

For speech recognition, three main forms of discriminative training have been examined, all of which can be expressed in terms of the posterior of the correct sentence. Using Bayes' rule this posterior may be expressed as

$$P(\boldsymbol{w}_{\texttt{ref}}^{(r)}|\boldsymbol{Y}^{(r)};\boldsymbol{\lambda}) = \frac{p(\boldsymbol{Y}^{(r)}|\boldsymbol{w}_{\texttt{ref}}^{(r)};\boldsymbol{\lambda})P(\boldsymbol{w}_{\texttt{ref}}^{(r)})}{\sum_{\boldsymbol{w}} p(\boldsymbol{Y}^{(r)}|\boldsymbol{w};\boldsymbol{\lambda})P(\boldsymbol{w})}, \qquad (4.2)$$

where the summation is over all possible word-sequences. By expressing the posterior in terms of the HMM generative model parameters, it is possible to estimate "generative" model parameters with discriminative criteria.

The language model (or class prior), $P(\boldsymbol{w})$, is not normally trained in conjunction with the acoustic model, (though there has been some work in this area [145]) since typically the amount of text training data for the language model is far greater (orders of magnitude) than the available acoustic training data. Note, however, that unlike the ML case, the language model is included in the denominator of the objective function in the discriminative training case.

In addition to the criteria discussed here, there has been work on maximum-margin based estimation schemes [79, 100, 156]. Also, interest is growing in using discriminative models for speech recognition, where $P(\boldsymbol{w}_{\mathrm{ref}}^{(r)}|\boldsymbol{Y}^{(r)};\boldsymbol{\lambda})$ is directly modelled [54, 67, 93, 95], rather than using Bayes' rule as above to obtain the posterior of the correct word sequence. In addition, these discriminative criteria can be used to train feature transforms [138, 197] and model parameter transforms [159] that are dependent on the observations making them time varying.

4.1.1 Maximum Mutual Information

One of the first discriminative training criteria to be explored was the maximum mutual information (MMI) criterion [9, 125]. Here, the aim is to maximise the mutual information between the word-sequence, $\boldsymbol{w}$, and the information extracted by a recogniser with parameters $\boldsymbol{\lambda}$ from the associated observation sequence, $\boldsymbol{Y}$, $\mathcal{I}(\boldsymbol{w},\boldsymbol{Y};\boldsymbol{\lambda})$. As the joint distribution of the word-sequences and observations is unknown, it is approximated by the empirical distributions over the training data. This can be expressed as [20]

$$\mathcal{I}(\boldsymbol{w},\boldsymbol{Y};\boldsymbol{\lambda}) \approx \frac{1}{R}\sum_{r=1}^{R}\log\left(\frac{P(\boldsymbol{w}_{\mathrm{ref}}^{(r)},\boldsymbol{Y}^{(r)};\boldsymbol{\lambda})}{P(\boldsymbol{w}_{\mathrm{ref}}^{(r)})p(\boldsymbol{Y}^{(r)};\boldsymbol{\lambda})}\right). \tag{4.3}$$

As only the acoustic model parameters are trained, $P(\boldsymbol{w}_{\mathrm{ref}}^{(r)})$ is fixed, this is equivalent to finding the model parameters that maximise the average log-posterior probability of the correct word sequence.[1] Thus

---

[1] Given that the class priors, the language model probabilities, are fixed this should really be called conditional entropy training. When this form of training criterion is used with discriminative models it is also known as conditional maximum likelihood (CML) training.

the following criterion is maximised

$$\mathcal{F}_{\texttt{mmi}}(\boldsymbol{\lambda}) = \frac{1}{R}\sum_{r=1}^{R}\log(P(\boldsymbol{w}_{\texttt{ref}}^{(r)}|\boldsymbol{Y}^{(r)};\boldsymbol{\lambda})) \tag{4.4}$$

$$= \frac{1}{R}\sum_{r=1}^{R}\log\left(\frac{p(\boldsymbol{Y}^{(r)}|\boldsymbol{w}_{\texttt{ref}}^{(r)};\boldsymbol{\lambda})P(\boldsymbol{w}_{\texttt{ref}}^{(r)})}{\sum_{\boldsymbol{w}}p(\boldsymbol{Y}^{(r)}|\boldsymbol{w};\boldsymbol{\lambda})P(\boldsymbol{w})}\right). \tag{4.5}$$

Intuitively, the numerator is the likelihood of the data given the correct word sequence $\boldsymbol{w}_{\texttt{ref}}^{(r)}$, whilst the denominator is the total likelihood of the data given all possible word sequences $\boldsymbol{w}$. Thus, the objective function is maximised by making the correct model sequence likely and all other model sequences unlikely.

4.1.2 Minimum Classification Error

Minimum classification error (MCE) is a smooth measure of the error [82, 29]. This is normally based on a smooth function of the difference between the log-likelihood of the correct word sequence and all other competing sequences, and a sigmoid is often used for this purpose. The MCE criterion may be expressed in terms of the posteriors as

$$\mathcal{F}_{\texttt{mce}}(\boldsymbol{\lambda}) = \frac{1}{R}\sum_{r=1}^{R}\left(1 + \left[\frac{P(\boldsymbol{w}_{\texttt{ref}}^{(r)}|\boldsymbol{Y}^{(r)};\boldsymbol{\lambda})}{\sum_{\boldsymbol{w}\neq\boldsymbol{w}_{\texttt{ref}}^{(r)}}P(\boldsymbol{w}|\boldsymbol{Y}^{(r)};\boldsymbol{\lambda})}\right]^{\varrho}\right)^{-1}. \tag{4.6}$$

There are some important differences between MCE and MMI. The first is that the denominator term does not include the correct word sequence. Secondly, the posteriors (or log-likelihoods) are smoothed with a sigmoid function, which introduces an additional smoothing term ϱ. When $\varrho = 1$ then

$$\mathcal{F}_{\texttt{mce}}(\boldsymbol{\lambda}) = 1 - \frac{1}{R}\sum_{r=1}^{R}P(\boldsymbol{w}_{\texttt{ref}}^{(r)}|\boldsymbol{Y}^{(r)};\boldsymbol{\lambda}). \tag{4.7}$$

This is a specific example of the minimum Bayes risk criterion discussed next.

4.1.3 Minimum Bayes' Risk

In minimum Bayes' risk (MBR) training, rather than trying to model the correct distribution, as in the MMI criterion, the expected loss during recognition is minimised [22, 84]. To approximate the expected loss during recognition, the expected loss estimated on the training data is used. Here

$$\mathcal{F}_{\text{mbr}}(\boldsymbol{\lambda}) = \frac{1}{R} \sum_{r=1}^{R} \sum_{\boldsymbol{w}} P(\boldsymbol{w}|\boldsymbol{Y}^{(r)}; \boldsymbol{\lambda}) \mathcal{L}(\boldsymbol{w}, \boldsymbol{w}_{\text{ref}}^{(r)}), \qquad (4.8)$$

where $\mathcal{L}(\boldsymbol{w}, \boldsymbol{w}_{\text{ref}}^{(r)})$ is the loss function of word sequence $\boldsymbol{w}$ against the reference for sequence r, $\boldsymbol{w}_{\text{ref}}^{(r)}$.

There are a number of loss functions that have been examined.

- *1/0 loss*: For continuous speech recognition this is equivalent to a sentence-level loss function

$$\mathcal{L}(\boldsymbol{w}, \boldsymbol{w}_{\text{ref}}^{(r)}) = \begin{cases} 1; & \boldsymbol{w} \neq \boldsymbol{w}_{\text{ref}}^{(r)} \\ 0; & \boldsymbol{w} = \boldsymbol{w}_{\text{ref}}^{(r)} \end{cases}.$$

  When $\varrho = 1$ MCE and MBR training with a sentence cost function are the same.
- *Word*: The loss function directly related to minimising the expected word error rate (WER). It is normally computed by minimising the Levenshtein edit distance.
- *Phone*: For large vocabulary speech recognition not all word sequences will be observed. To assist generalisation, the loss function is often computed between phone sequences, rather than word sequences. In the literature this is known as minimum phone error (MPE) training [136, 139].
- *Phone frame error*: When using the phone loss-function, the number of possible errors to be corrected is reduced compared to the number of frames. This can cause generalisation issues. To address this minimum phone frame error (MPFE) may be used where the phone loss is weighted by the number of frames associated with each frame [198]. This is the same as the Hamming distance described in [166].

It is also possible to base the loss function on the specific task for which the classifier is being built [22].

A comparison of the MMI, MCE and MPE criteria on the Wall Street Journal (WSJ) task and a general framework for discriminative training is given in [111]. Though all the criteria significantly outperformed ML training, MCE and MPE were found to outperform MMI on this task. In [198], MPFE is shown to give small but consistent gains over MPE.

4.2   Implementation Issues

In *Architecture of an HMM-Based Recogniser*, the use of the EM algorithm to train the parameters of an HMM using the ML criterion was described. This section briefly discusses some of the implementation issues that need to be addressed when using discriminative training criteria. Two aspects are considered. Firstly, the form of algorithm used to estimate the model parameters is described. Secondly, since these discriminative criteria can over-train, various techniques for improving generalisation are detailed.

4.2.1   Parameter Estimation

EM cannot be used to estimate parameters when discriminative criteria are used. To see why this is the case, note that the MMI criterion can be expressed as the difference of two log-likelihood expressions

$$\mathcal{F}_{\texttt{mmi}}(\boldsymbol{\lambda}) = \frac{1}{R}\sum_{r=1}^{R}\left(\log\left(p(\boldsymbol{Y}^{(r)}|\boldsymbol{w}_{\texttt{ref}}^{(r)};\boldsymbol{\lambda})P(\boldsymbol{w}_{\texttt{ref}}^{(r)}) \right) \right.$$

$$\left. - \log\left(\sum_{\boldsymbol{w}} p(\boldsymbol{Y}^{(r)}|\boldsymbol{w};\boldsymbol{\lambda})P(\boldsymbol{w}) \right) \right). \quad (4.9)$$

The first term is referred to as the *numerator term* and the second, the *denominator term*. The numerator term is identical to the standard ML criterion (the language model term, $P(\boldsymbol{w}_{\texttt{ref}}^{(r)})$, does not influence the ML-estimate). The denominator may also be rewritten in a similar fashion to the numerator by producing a composite HMM with parameters $\boldsymbol{\lambda}^{\texttt{den}}$. Hence, although auxiliary functions may be computed for each of these, the difference of two lower-bounds is not itself a lower-bound and so standard EM cannot be used. To handle this problem, the extended Baum-Welch (EBW) criterion was proposed [64, 129]. In this case, standard EM-like auxiliary functions are defined for the numerator and denominator but stability during re-estimation is achieved by adding scaled current model parameters to the numerator statistics. The weighting of these parameters is specified by a constant D. A large enough value of D can be shown to guarantee that the criterion does not decrease. For MMI, the update of the mean parameters then becomes

$$\hat{\boldsymbol{\mu}}^{(jm)} = \frac{\sum_{r=1}^{R}\sum_{t=1}^{T^{(r)}}\left((\gamma_{\mathsf{numt}}^{(rjm)} - \gamma_{\mathsf{dent}}^{(rjm)})\boldsymbol{y}_{t}^{(r)}\right) + D\boldsymbol{\mu}^{(jm)}}{\sum_{r=1}^{R}\sum_{t=1}^{T^{(r)}}\left(\gamma_{\mathsf{numt}}^{(rjm)} - \gamma_{\mathsf{dent}}^{(rjm)}\right) + D}, \qquad (4.10)$$

where $\boldsymbol{\mu}^{(jm)}$ are the current model parameters, $\gamma_{\mathsf{numt}}^{(rm)}$ is the same as the posterior used in ML estimation and

$$\gamma_{\mathsf{dent}}^{(rjm)} = P(\theta_t = \mathsf{s}_{jm}|\boldsymbol{Y}^{(r)};\boldsymbol{\lambda}^{\mathsf{den}}) \qquad (4.11)$$

and $\boldsymbol{\lambda}^{\mathsf{den}}$ is the composite HMM representing all competing paths. One issue with EBW is the speed of convergence, which may be very slow. This may be made faster by making D Gaussian component specific [136]. A similar expression can be derived from a *weak-sense* auxiliary function perspective [136]. Here an auxiliary function is defined which rather than being a strict lower-bound, simply has the same gradient as the criterion at the current model parameters. Similar expressions can be derived for the MPE criterion, and recently large vocabulary training with MCE has also been implemented [119].

For large vocabulary speech recognition systems, where thousands of hours of training data may be used, it is important that the training procedure be as efficient as possible. For discriminative techniques one of the major costs is the accumulation of the statistics for the denominator since this computation is equivalent to recognising the training data. To improve efficiency, it is common to use lattices as a compact representation of the most likely competing word sequences and only accumulate statistics at each iteration for paths in the lattice [174].

4.2.2 Generalisation

Compared to ML training, it has been observed that discriminative criteria tend to generalise less-well. To mitigate this, a number of techniques have been developed that improve the robustness of the estimates.

As a consequence of the conditional independence assumptions, the posterior probabilities computed using the HMM likelihoods tend to have a very large dynamic range and typically one of the hypotheses dominates. To address this problem, the acoustic model likelihoods are often raised to a fractional power, referred to as acoustic deweighting [182]. Thus when accumulating the statistics, the posteriors are based on

$$P(\boldsymbol{w}_{\mathrm{ref}}^{(r)}|\boldsymbol{Y}^{(r)};\boldsymbol{\lambda}) = \frac{p(\boldsymbol{Y}^{(r)}|\boldsymbol{w}_{\mathrm{ref}}^{(r)};\boldsymbol{\lambda})^{\alpha}P(\boldsymbol{w}_{\mathrm{ref}}^{(r)})^{\beta}}{\sum_{\boldsymbol{w}}p(\boldsymbol{Y}^{(r)}|\boldsymbol{w};\boldsymbol{\lambda})^{\alpha}P(\boldsymbol{w})^{\beta}}. \qquad (4.12)$$

In practice β is often set to one and α is set to the inverse of the language model scale-factor described in footnote 2 in *Architecture of an HMM-Based Recogniser*.

The form of the language model used in training should in theory match the form used for recognition. However, it has been found that using simpler models, unigrams or heavily pruned bigrams, for training despite using trigrams or fourgrams in decoding improves performance [152]. By weakening the language model, the number of possible confusions is increased allowing more complex models to be trained given a fixed quantity of training data.

To improve generalisation, "robust" parameter priors may be used when estimating the models. These priors may either be based on the ML parameter estimates [139] or, for example when using MPE training, the MMI estimates [150]. For MPE training using these priors has been found to be essential for achieving performance gains [139].

4.3 Lightly Supervised and Unsupervised Training

One of the major costs in building an acoustic training corpus is the production of accurate orthographic transcriptions. For some situations, such as broadcast news (BN) which has associated closed-captions, approximate transcriptions may be derived. These transcriptions are known to be error-full and thus not suitable for direct use when training detailed acoustic models. However, a number of *lightly supervised* training techniques have been developed to overcome this [23, 94, 128].

The general approach to lightly supervised training is exemplified by the procedure commonly used for training on BN data using closed-captions (CC). This procedure consists of the following stages:

(1) Construct a language model (LM) using only the CC data. This CC LM is then interpolated with a general BN LM using interpolation weights heavily weighted towards the CC LM. For example in [56] the interpolation weights were 0.9 and 0.1.[2] This yields a *biased* language model.

(2) Recognise the audio data using an existing acoustic model and the biased LM trained in (1).

(3) Optionally post-process the data. For example, only use segments from the training data where the recognition output from (2) is consistent to some degree with the CCs, or only use segments with high confidence in the recognition output.

(4) Use the selected segments for acoustic model training with the hypothesised transcriptions from (2).

For acoustic data which has no transcriptions at all, unsupervised training techniques must be used [86, 110]. Unsupervised training is similar to lightly supervised training except that a general recognition language model is used, rather than the biased language model for lightly supervised training. This reduces the accuracy of the transcriptions [23], but has been successfully applied to a range of tasks [56, 110]. However, the gains from unsupervised approaches can decrease dramatically when discriminative training such as MPE is used. If the mismatch between the supervised training data and unsupervised data is

---

[2] The interpolation weights were found to be relatively insensitive to the type of source data when tuned on held-out data.

large, the accuracy of the numerator transcriptions become very poor. The gains obtained from using the un-annotated data then become small [179]. Thus currently there appears to be a limit to how poor the transcriptions can be whilst still obtaining worthwhile gains from discriminative training. This problem can be partly overcome by using the recognition output to guide the selection of data to manually transcribe [195].

5

Adaptation and Normalisation

A fundamental idea in statistical pattern classification is that the training data should adequately represent the test data, otherwise a mismatch will occur and recognition accuracy will be degraded. In the case of speech recognition, there will always be new speakers who are poorly represented by the training data, and new hitherto unseen environments. The solution to these problems is *adaptation*. Adaptation allows a small amount of data from a target speaker to be used to transform an acoustic model set to make it more closely match that speaker. It can be used both in training to make more specific or more compact recognition sets, and it can be used in recognition to reduce mismatch and the consequent recognition errors.

There are varying styles of adaptation which affect both the possible applications and the method of implementation. Firstly, adaptation can be *supervised* in which case accurate transcriptions are available for all of the adaptation data, or it can be *unsupervised* in which case the required transcriptions must be hypothesised. Secondly, adaptation can be *incremental* or *batch-mode*. In the former case, adaptation data becomes available in stages, for example, as is the case for a spoken dialogue system, when a new caller comes on the line. In batch-mode, all of the adaptation data is available from the start as is the case in off-line transcription.

A range of adaptation and normalisation approaches have been proposed. This section gives an overview of some of the more common schemes. First feature-based approaches, which only depend on the acoustic features, are described. Then *maximum a posteriori* adaptation and cluster based approaches are described. Finally, linear transformation-based approaches are discussed. Many of these approaches can be used within an *adaptive training* framework which is the final topic.

5.1   Feature-Based Schemes

Feature-based schemes only depend on the acoustic features. This is only strictly true of the first two approaches: mean and variance normalisation; and Gaussianisation. The final approach, vocal tract length normalisation can be implemented in a range of ways, many of which do not fully fit within this definition.

5.1.1   Mean and Variance Normalisation

Cepstral mean normalisation (CMN) removes the average feature value of the feature-vector from each observation. Since cepstra in common with most front-end feature sets are derived from log spectra, this has the effect of reducing sensitivity to channel variation, provided that the channel effects do not significantly vary with time or the amplitude of the signal. Cepstral variance normalisation (CVN) scales each individual feature coefficient to have a unit variance and empirically this has been found to reduce sensitivity to additive noise [71].

For transcription applications where multiple passes over the data are possible, the necessary mean and variance statistics should be computed over the longest possible segment of speech for which the speaker and environment conditions are constant. For example, in BN transcription this will be a speaker segment and in telephone transcription it will be a whole side of a conversation. Note that for real time systems which operate in a single continuous pass over the data, the mean and variance statistics must be computed as running averages.

5.1.2   Gaussianisation

Given that normalising the first and second-order statistics yields improved performance, an obvious extension is to normalise the higher order statistics for the data from, for example, a specific speaker so that the overall distribution of all the features from that speaker is Gaussian. This so-called *Gaussianisation* is performed by finding a transform $\boldsymbol{y} = \phi(\tilde{\boldsymbol{y}})$, that yields

$$\boldsymbol{y} \sim \mathcal{N}(\mathbf{0}, \mathbf{I}), \tag{5.1}$$

where $\mathbf{0}$ is the d-dimensional zero vector and $\mathbf{I}$ is a d-dimensional identity matrix. In general performing Gaussianisation on the complete feature vector is highly complex. However, if each element of the feature vector is treated independently, then there a number of schemes that may be used to define $\phi(\cdot)$.

One standard approach is to estimate a cumulative distribution function (CDF) for each feature element $\tilde{y}_i$, $F(\tilde{y}_i)$. This can be estimated using all the data points from, for example, a speaker in a histogram approach. Thus for dimension i of base observation $\tilde{\boldsymbol{y}}$

$$F(\tilde{y}_i) \approx \frac{1}{T} \sum_{t=1}^{T} h(\tilde{y}_i - \tilde{y}_{ti}) = \mathrm{rank}(\tilde{y}_i)/T, \tag{5.2}$$

where $\tilde{\boldsymbol{y}}_1, \ldots, \tilde{\boldsymbol{y}}_T$ are the data, $h(\cdot)$ is the step function and $\mathrm{rank}(\tilde{y}_i)$ is the rank of element i of $\tilde{\boldsymbol{y}}$, $\tilde{y}_i$, when the data are sorted. The transformation required such that the transformed dimension i of observation $\boldsymbol{y}$ is approximately Gaussian is

$$y_i = \Phi^{-1}\left(\frac{\mathrm{rank}(\tilde{y}_i)}{T}\right), \tag{5.3}$$

where $\Phi(\cdot)$ is the CDF of a Gaussian [148]. Note, however, that although each dimension will be approximately Gaussian distributed, the complete feature vector will not necessarily satisfy (5.1) since the elements may be correlated with one another.

One difficulty with this approach is that when the normalisation data set is small the CDF estimate, $F(\tilde{y}_i)$, can be noisy. An alternative approach is to estimate an M-component GMM on the data and then use this to approximate the CDF [26], that is

$$y_i = \Phi^{-1}\left(\int_{-\infty}^{\tilde{y}_i} \sum_{m=1}^{M} c_m \mathcal{N}(z; \mu_i^{(m)}, \sigma_i^{(m)2}) dz\right), \qquad (5.4)$$

where $\mu_i^{(m)}$ and $\sigma_i^{(m)2}$ are the mean and variance of the mth GMM component of dimension i trained on all the data. This results in a smoother and more compact representation of the Gaussianisation transformation [55].

Both these forms of Gaussianisation have assumed that the dimensions are independent of one another. To reduce the impact of this assumption Gaussianisation is often applied after some form of decorrelating transform. For example, a global semi-tied transform can be applied prior to the Gaussianisation process.

5.1.3 Vocal Tract Length Normalisation

Variations in vocal tract length cause formant frequencies to shift in frequency in an approximately linear fashion. Thus, one simple form of normalisation is to linearly scale the filter bank centre frequencies within the front-end feature extractor to approximate a canonical formant frequency scaling [96]. This is called *vocal tract length normalisation (VTLN)*.

To implement VTLN two issues need to be addressed: definition of the scaling function and estimation of the appropriate scaling function parameters for each speaker. Early attempts at VTLN used a simple linear mapping, but as shown in Figure 5.1(a) this results in a problem at high frequencies where female voices have no information in the upper frequency band and male voices have the upper frequency band truncated. This can be mitigated by using a piece-wise linear function of the form shown in Figure 5.1(b) [71]. Alternatively, a bilinear transform can be used [120]. Parameter estimation is performed using a grid search plotting log likelihoods against parameter values. Once the optimal values for all training speakers have been computed, the training data is normalised and the acoustic models re-estimated. This is repeated until the VTLN parameters have stabilised. Note here that

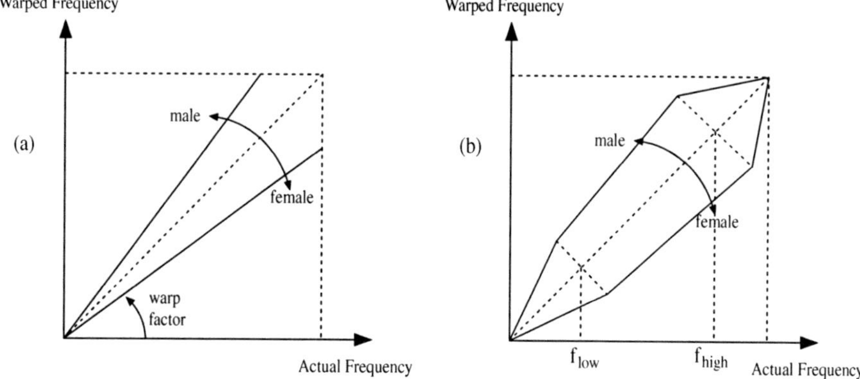

Fig. 5.1 Vocal tract length normalisation.

when comparing log likelihoods resulting from differing VTLN transformations, then the Jacobian of the transform should be included. This is however very complex to estimate and since the application of mean and variance normalisation will reduce the affect of this approximation, it is usually ignored.

For very large systems, the overhead incurred from iteratively computing the optimal VTLN parameters can be considerable. An alternative is to approximate the effect of VTLN by a linear transform. The advantage of this approach is that the optimal transformation parameters can be determined from the auxiliary function in a single pass over the data [88].

VTLN is particularly effective for telephone speech where speakers can be clearly identified. It is less effective for other applications such as broadcast news transcription where speaker changes must be inferred from the data.

5.2 Linear Transform-Based Schemes

For cases where adaptation data is limited, linear transformation based schemes are currently the most effective form of adaptation. These differ from feature-based approaches in that they use the acoustic model parameters and they require a transcription of the adaptation data.

5.2.1 Maximum Likelihood Linear Regression

In *maximum likelihood linear regression (MLLR)*, a set of linear transforms are used to map an existing model set into a new adapted model set such that the likelihood of the adaptation data is maximised. MLLR is generally very robust and well suited to unsupervised incremental adaptation. This section presents MLLR in the form of a single global linear transform for all Gaussian components. The multiple transform case, where different transforms are used depending on the Gaussian component to be adapted, is discussed later.

There are two main variants of MLLR: unconstrained and constrained [50, 97]. In unconstrained MLLR, separate transforms are trained for the means and variances

$$\hat{\boldsymbol{\mu}}^{(sm)} = \mathbf{A}^{(s)}\boldsymbol{\mu}^{(m)} + \mathbf{b}^{(s)}; \quad \hat{\boldsymbol{\Sigma}}^{(sm)} = \mathbf{H}^{(s)}\boldsymbol{\Sigma}^{(m)}\mathbf{H}^{(s)\mathsf{T}} \tag{5.5}$$

where s indicates the speaker. Although (5.5) suggests that the likelihood calculation is expensive to compute, unless $\mathbf{H}$ is constrained to be diagonal, it can in fact be made efficient using the following equality

$$\mathcal{N}\left(\boldsymbol{y}; \hat{\boldsymbol{\mu}}^{(sm)}, \hat{\boldsymbol{\Sigma}}^{(sm)}\right)$$

$$= \frac{1}{|\mathbf{H}^{(s)}|}\mathcal{N}(\mathbf{H}^{(s)-1}\boldsymbol{y}; \mathbf{H}^{(s)-1}\mathbf{A}^{(s)}\boldsymbol{\mu}^{(m)} + \mathbf{H}^{(s)-1}\mathbf{b}^{(s)}, \boldsymbol{\Sigma}^{(m)}). \tag{5.6}$$

If the original covariances are diagonal (as is common), then by appropriately caching the transformed observations and means, the likelihood can be calculated at the same cost as when using the original diagonal covariance matrices.

For MLLR there are no constraints between the adaptation applied to the means and the covariances. If the two matrix transforms are constrained to be the same, then a linear transform related to the feature-space transforms described earlier may be obtained. This is constrained MLLR (CMLLR)

$$\hat{\boldsymbol{\mu}}^{(sm)} = \tilde{\mathbf{A}}^{(s)}\boldsymbol{\mu}^{(m)} + \tilde{\mathbf{b}}^{(s)}; \quad \hat{\boldsymbol{\Sigma}}^{(sm)} = \tilde{\mathbf{A}}^{(s)}\boldsymbol{\Sigma}^{(m)}\tilde{\mathbf{A}}^{(s)\mathsf{T}}. \tag{5.7}$$

In this case, the likelihood can be expressed as

$$\mathcal{N}(\boldsymbol{y}; \hat{\boldsymbol{\mu}}^{(sm)}, \hat{\boldsymbol{\Sigma}}^{(sm)}) = |\mathbf{A}^{(s)}|\mathcal{N}(\mathbf{A}^{(s)}\boldsymbol{y} + \mathbf{b}^{(s)}; \boldsymbol{\mu}^{(m)}, \boldsymbol{\Sigma}^{(m)}), \tag{5.8}$$

where

$$\mathbf{A}^{(s)} = \tilde{\mathbf{A}}^{(s)-1}; \quad \mathbf{b}^{(s)} = -\tilde{\mathbf{A}}^{(s)-1}\tilde{\mathbf{b}}^{(s)}. \tag{5.9}$$

Thus with this constraint, the actual model parameters are not transformed making this form of transform efficient if the speaker (or acoustic environment) changes rapidly. CMLLR is the form of linear transform most often used for adaptive training, discussed later.

For both forms of linear transform, the matrix transformation may be full, block-diagonal, or diagonal. For a given amount of adaptation, more diagonal transforms may be reliably estimated than full ones. However, in practice, full transforms normally outperform larger numbers of diagonal transforms [126]. Hierarchies of transforms of different complexities may also be used [36].

5.2.2 Parameter Estimation

The maximum likelihood estimation formulae for the various forms of linear transform are given in [50, 97]. Whereas there are closed-form solutions for unconstrained mean MLLR, the constrained and unconstrained variance cases are similar to the semi-tied covariance transform discussed in *HMM Structure Refinements* and they require an iterative solution.

Both forms of linear transforms require transcriptions of the adaptation data in order to estimate the model parameters. For supervised adaptation, the transcription is known and may be directly used without further consideration. When used in unsupervised mode, the transcription must be derived from the recogniser output and in this case, MLLR is normally applied iteratively [183] to ensure that the best hypothesis for estimating the transform parameters is used. First, the unknown speech is recognised, then the hypothesised transcription is used to estimate MLLR transforms. The unknown speech is then re-recognised using the adapted models. This is repeated until convergence is achieved.

Using this approach, all words within the hypothesis are treated as equally probable. A refinement is to use recognition lattices in place of the 1-best hypothesis to accumulate the adaptation statistics. This approach is more robust to recognition errors and avoids the need to re-recognise the data since the lattice can be simply rescored [133]. An alternative use of lattices is to obtain confidence scores, which may then be used for *confidence-based* MLLR [172]. A different approach using N-best lists was proposed in [118, 194] whereby a separate transform was estimated for each hypothesis and used to rescore only that hypothesis. In [194], this is shown to be a lower-bound approximation where the transform parameters are considered as latent variables.

The initial development of transform-based adaptation methods used the ML criterion, this was then extended to include maximum *a posteriori* estimation [28]. Linear transforms can also be estimated using discriminative criteria [171, 178, 180]. For supervised adaptation, any of the standard approaches may be used. However, if unsupervised adaptation is used, for example in BN transcription systems, then there is an additional concern. As discriminative training schemes attempt to modify the parameters so that the posterior of the transcription (or a function thereof) is improved, it is more sensitive to errors in the transcription hypotheses than ML estimation. This is the same issue as was observed for unsupervised discriminative training and, in practice, discriminative unsupervised adaptation is not commonly used.

5.2.3 Regression Class Trees

A powerful feature of linear transform-based adaptation is that it allows all the acoustic models to be adapted using a variable number of transforms. When the amount of adaptation data is limited, a global transform can be shared across all Gaussians in the system. As the amount of data increases, the HMM state components can be grouped into regression classes with each class having its own transform, for example $\mathbf{A}^{(r)}$ for regression class r. As the amount of data increases further, the number of classes and therefore transforms can be increased correspondingly to give better and better adaptation [97].

The number of transforms to use for any specific set of adaptation data can be determined automatically using a *regression class tree*

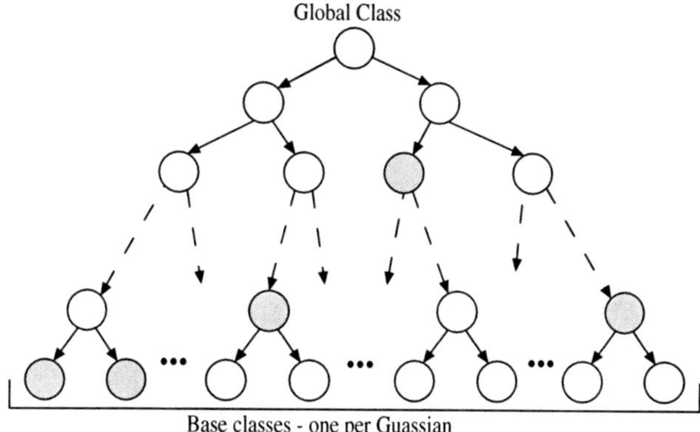

Fig. 5.2 A regression class tree.

as illustrated in Figure 5.2. Each node represents a regression class, i.e., a set of Gaussian components which will share a single transform. The total occupation count associated with any node in the tree can easily be computed since the counts are known at the leaf nodes. Then, for a given set of adaptation data, the tree is descended and the most specific set of nodes is selected for which there is sufficient data (for example, the shaded nodes in Figure 5.2). Regression class trees may either be specified using expert knowledge, or more commonly by automatically training the tree by assuming that Gaussian components that are "close" to one another are transformed using the same linear transform [97].

5.3 Gender/Cluster Dependent Models

The models described above have treated all the training data as a single-block, with no information about the individual segments of training data. In practice it is possible to record, or infer, information about the segment, for example the gender of the speaker. If this gender information is then used in training, the result is referred to as a *gender-dependent (GD)* system. One issue with training these GD models is that during recognition the gender of the test-speaker must be determined. As this is a single binary latent variable, it can normally be rapidly and robustly estimated.

It is also possible to extend the number of clusters beyond simple gender dependent models. For example speakers can be automatically clustered together to form clusters [60].

Rather than considering a single cluster to represent the speaker (and possibly acoustic environment), multiple clusters or so-called *eigenvoices* may be combined together. There are two approaches that have been adopted to combining the clusters together: likelihood combination; and parameter combination. In addition to combining the clusters, the form of representation of the cluster must be decided.

The simplest approach is to use a mixture of eigenvoices. Here

$$p(\boldsymbol{Y}_{1:T}|\boldsymbol{\nu}^{(s)};\boldsymbol{\lambda}) = \sum_{k=1}^{K} \nu_k^{(s)} p(\boldsymbol{Y}_{1:T};\boldsymbol{\lambda}^{(k)}), \tag{5.10}$$

where $\boldsymbol{\lambda}^{(k)}$ are the model parameters associated with the kth cluster, s indicates the speaker, and

$$\sum_{k=1}^{K} \nu_k^{(s)} = 1, \quad \nu_k^{(s)} \geq 0. \tag{5.11}$$

Thus gender dependent models are a specific case where $\nu_k^{(s)} = 1$ for the "correct" gender and zero otherwise. One problem with this approach is that all K cluster likelihoods must be computed for each observation sequence. The training of these forms of model can be implemented using EM. This form of interpolation can also be combined with linear transforms [35].

Rather than interpolating likelihoods, the model parameters themselves can be interpolated. Thus

$$p(\boldsymbol{Y}_{1:T}|\boldsymbol{\nu}^{(s)};\boldsymbol{\lambda}) = p\left(\boldsymbol{Y}_{1:T}; \sum_{k=1}^{K} \nu_k^{(s)} \boldsymbol{\lambda}^{(k)}\right). \tag{5.12}$$

Current approaches limit interpolation to the mean parameters, and the remaining variances, priors and transition probabilities are tied over all the clusters

$$\hat{\boldsymbol{\mu}}^{(sm)} = \sum_{k=1}^{K} \nu_k^{(s)} \boldsymbol{\mu}^{(km)}. \tag{5.13}$$

Various forms of this model have been investigated: reference speaker weighting (RSW) [73], EigenVoices [91], and cluster adaptive training (CAT) [52]. All use the same interpolation of the means, but differ in how the eigenvoices are estimated and how the interpolation weight vector, $\nu^{(s)}$, is estimated. RSW is the simplest approach using each speaker as an eigenvoice. In the original implementation of EigenVoices, the cluster means were determined using PCA on the extended mean super-vector, where all the means of the models for each speaker were concatenated together. In contrast, CAT uses ML to estimate the cluster parameters. Note that CAT parameters are estimated using an adaptive training style which is discussed in more detail below.

Having estimated the cluster parameters in training, for the test speaker the interpolation weights, $\nu^{(s)}$, are estimated. For both Eigen-Voices and CAT, the ML-based smoothing approach in [49] is used. In contrast RSW uses ML with the added constraint that the weights satisfy the constraints in (5.11). This complicates the optimisation criterion and hence it is not often used.

A number of extensions to this basic parameter interpolation framework have been proposed. In [52], linear transforms were used to represent each cluster for CAT, also referred to as Eigen-MLLR [24] when used in an Eigenvoice-style approach. Kernelised versions of both Eigen-Voices [113] and Eigen-MLLR [112] have also been studied. Finally, the use of discriminative criteria to obtain the cluster parameters has been derived [193].

5.4   Maximum a Posteriori (MAP) Adaptation

Rather than hypothesising a form of transformation to represent the differences between speakers, it is possible to use standard statistical approaches to obtain robust parameter estimates. One common approach is maximum *a posteriori* (MAP) adaptation where in addition to the adaptation data, a prior over the model parameters, $p(\boldsymbol{\lambda})$, is used to estimate the model parameters. Given some adaptation data $\boldsymbol{Y}^{(1)}, \ldots, \boldsymbol{Y}^{(R)}$ and a model with parameters $\boldsymbol{\lambda}$, MAP-based parameter estimation seeks to maximise the following objective function

$$\mathcal{F}_{\tt map}(\boldsymbol{\lambda}) = \mathcal{F}_{\tt ml}(\boldsymbol{\lambda}) + \frac{1}{R}\log(p(\boldsymbol{\lambda}))$$

$$= \left(\frac{1}{R}\sum_{r=1}^{R}\log p(\boldsymbol{Y}^{(r)}|\boldsymbol{w}_{\tt ref}^{(r)};\boldsymbol{\lambda}) \right) + \frac{1}{R}\log(p(\boldsymbol{\lambda})). \qquad (5.14)$$

Comparing this with the ML objective function given in (4.1), it can be seen that the likelihood is weighted by the prior. Note that MAP can also be used with other training criteria such as MPE [137]. The choice of distribution for this prior is problematic since there is no conjugate prior density for a continuous Gaussian mixture HMM. However, if the mixture weights and Gaussian component parameters are assumed to be independent, then a finite mixture density of the form $p_D(\boldsymbol{c}_j)\prod_m p_W(\boldsymbol{\mu}^{(jm)},\boldsymbol{\Sigma}^{(jm)})$ can be used where $p_D(\cdot)$ is a Dirichlet distribution over the vector of mixture weights $\boldsymbol{c}_j$ and $p_W(\cdot)$ is a normal-Wishart density. It can then be shown that this leads to parameter estimation formulae of the form [61]

$$\hat{\boldsymbol{\mu}}^{(jm)} = \frac{\tau\boldsymbol{\mu}^{(jm)} + \sum_{r=1}^{R}\sum_{t=1}^{T^{(r)}}\gamma_t^{(rjm)}\boldsymbol{y}_t^{(r)}}{\tau + \sum_{r=1}^{R}\sum_{t=1}^{T^{(r)}}\gamma_t^{(rjm)}}, \qquad (5.15)$$

where $\boldsymbol{\mu}^{(jm)}$ is the prior mean and τ is a parameter of $p_W(\cdot)$ which is normally determined empirically. Similar, though rather more complex, formulae can be derived for the variances and mixture weights [61].

Comparing (5.15) with (2.13), it can be seen that MAP adaptation effectively interpolates the original prior parameter values with those that would be obtained from the adaptation data alone. As the amount of adaptation data increases, the parameters tend asymptotically to the adaptation domain. This is a desirable property and it makes MAP especially useful for porting a well-trained model set to a new domain for which there is only a limited amount of data.

A major drawback of MAP adaptation is that every Gaussian component is updated individually. If the adaptation data is sparse, then many of the model parameters will not be updated. Various attempts have been made to overcome this (e.g. [4, 157]) but MAP nevertheless remains ill-suited for rapid incremental adaptation.

5.5 Adaptive Training

Ideally, an acoustic model set should encode just those dimensions which allow the different classes to be discriminated. However, in the case of speaker independent (SI) speech recognition, the training data necessarily includes a large number of speakers. Hence, acoustic models trained directly on this set will have to "waste" a large number of parameters encoding the variability between speakers rather than the variability between spoken words which is the true aim. One approach to handling this problem is to use adaptation transforms during training. This is referred to as speaker adaptive training (SAT) [5].

The concept behind SAT is illustrated in Figure 5.3. For each training speaker, a transform is estimated, and then the canonical model is estimated given all of these speaker transforms. The complexity of performing adaptive training depends on the nature of the adaptation/normalisation transform. These may be split into three groups.

- *Model independent*: These schemes do not make explicit use of any model information. CMN/CVN and Gaussianisation belong to this group. VTLN may be independent of the model for example if estimated using attributes of the signal [96]. In these cases the features are transformed and the models trained as usual.

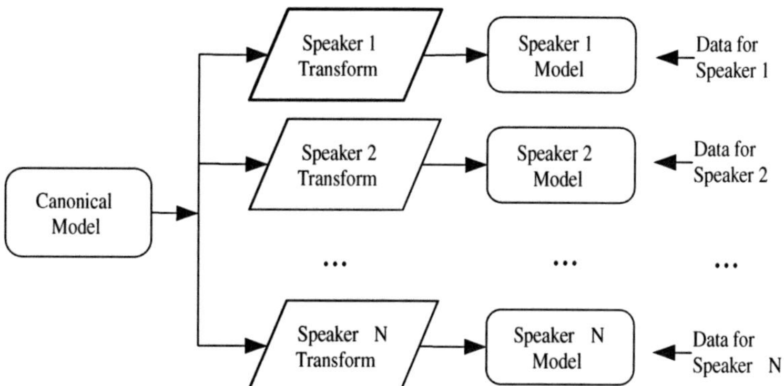

Fig. 5.3 Adaptive training.

- *Feature transformation*: These transforms also act on the features but are derived, normally using ML estimation, using the current estimate of the model set. Common versions of these feature transforms are VTLN using ML-style estimation, or the linear approximation [88] and constrained MLLR [50].
- *Model transformation*: The model parameters, means and possibly variances, are transformed. Common schemes are SAT using MLLR [5] and CAT [52].

For schemes where the transforms are dependent on the model parameters, the estimation is performed by an iterative process:

(1) initialise the canonical model using a speaker-independent system and set the transforms for each speaker to be an identity transform ($\mathbf{A}^{(s)} = \mathbf{I}, \mathbf{b}^{(s)} = \mathbf{0}$) for each speaker;
(2) estimate a transform for each speaker using only the data for that speaker;
(3) estimate the canonical model given each of the speaker transforms;
(4) Goto (2) unless converged or some maximum number of iterations reached.

When MLLR is used, the estimation of the canonical model is more expensive [5] than standard SI training. This is not the case for CAT, although additional parameters must be stored. The most common version of adaptive training uses CMLLR, since it is the simplest to implement. Using CMLLR, the canonical mean is estimated by

$$\hat{\boldsymbol{\mu}}^{(jm)} = \frac{\sum_{s=1}^{S} \sum_{t=1}^{T^{(s)}} \gamma_t^{(sjm)} \left(\mathbf{A}^{(s)} \boldsymbol{y}_t^{(s)} + \mathbf{b}^{(s)} \right)}{\sum_{s=1}^{S} \sum_{t=1}^{T^{(s)}} \gamma_t^{(sjm)}}, \qquad (5.16)$$

where S denotes the number of speakers and $\gamma_t^{(sjm)}$ is the posterior occupation probability of component m of state $\mathbf{s}_j$ generating the observation at time t based on data for speaker s. In practice, multiple forms of adaptive training can be combined together, for example using Gaussianisation with CMLLR-based SAT [55].

Finally, note that SAT trained systems incur the problem that they can only be used once transforms have been estimated for the test data. Thus, an SI model set is typically retained to generate the initial hypothesised transcription or lattice needed to compute the first set of transforms. Recently a scheme for dealing with this problem using transform priors has been proposed [194].

6

Noise Robustness

In *Adaptation and Normalisation*, a variety of schemes were described for adapting a speech recognition system to changes in speaker and environment. One very common and often extreme form of environment change is caused by ambient noise and although the general adaptation techniques described so far can reduce the effects of noise, specific noise compensation algorithms can be more effective, especially with very limited adaptation data.

Noise compensation algorithms typically assume that a clean speech signal x_t in the time domain is corrupted by additive noise n_t and a stationary convolutive channel impulse response h_t to give the observed noisy signal y_t [1]. That is

$$y_t = x_t \otimes h_t + n_t, \tag{6.1}$$

where $\otimes$ denotes convolution.

In practice, as explained in *Architecture of an HMM-Based Recogniser*, speech recognisers operate on feature vectors derived from a transformed version of the log spectrum. Thus, for example, in the mel-cepstral domain, (6.1) leads to the following relationship between the static clean speech, noise and noise corrupted speech observations[1]

$$\begin{aligned}
\boldsymbol{y}_t^{\mathrm{s}} &= \boldsymbol{x}_t^{\mathrm{s}} + \boldsymbol{h} + \mathbf{C}\log\left(1 + \exp(\mathbf{C}^{-1}(\boldsymbol{n}_t^{\mathrm{s}} - \boldsymbol{x}_t^{\mathrm{s}} - \boldsymbol{h}))\right) \\
&= \boldsymbol{x}_t^{\mathrm{s}} + f(\boldsymbol{x}_t^{\mathrm{s}}, \boldsymbol{n}_t^{\mathrm{s}}, \boldsymbol{h}),
\end{aligned} \tag{6.2}$$

where $\mathbf{C}$ is the DCT matrix. For a given set of noise conditions, the observed (static) speech vector $\boldsymbol{y}_t^{\mathrm{s}}$ is a highly non-linear function of the underlying clean (static) speech signal $\boldsymbol{x}_t^{\mathrm{s}}$. In broad terms, a small amount of additive Gaussian noise adds a further mode to the speech distribution. As the signal to noise ratio falls, the noise mode starts to dominate, the noisy speech means move towards the noise and the variances shrink. Eventually, the noise dominates and there is little or no information left in the signal. The net effect is that the probability distributions trained on clean data become very poor estimates of the data distributions observed in the noisy environment and recognition error rates increase rapidly.

Of course, the ideal solution to this problem would be to design a feature representation which is immune to noise. For example, PLP features do exhibit some measure of robustness [74] and further robustness can be obtained by bandpass filtering the features over time in a process called relative spectral filtering to give so-called RASTA-PLP coefficients [75]. Robustness to noise can also be improved by training on speech recorded in a variety of noise conditions, this is called *multi-style* training. However, there is a limit to the effectiveness of these approaches and in practice SNRs below 15 dB or so require active noise compensation to maintain performance.

There are two main approaches to achieving noise robust recognition and these are illustrated in Figure 6.1 which shows the separate clean training and noisy recognition environments. Firstly, in *feature compensation*, the noisy features are compensated or enhanced to remove the effects of the noise. Recognition and training then takes place in the clean environment. Secondly, in *model compensation*, the clean acoustic models are compensated to match the noisy environment. In this latter case, training takes place in the clean environment, but the clean

---

[1] In this section $\log(\cdot)$ and $\exp(\cdot)$ when applied to a vector perform an element-wise logarithm or exponential function to all elements of the vector. They will thus yield a vector.

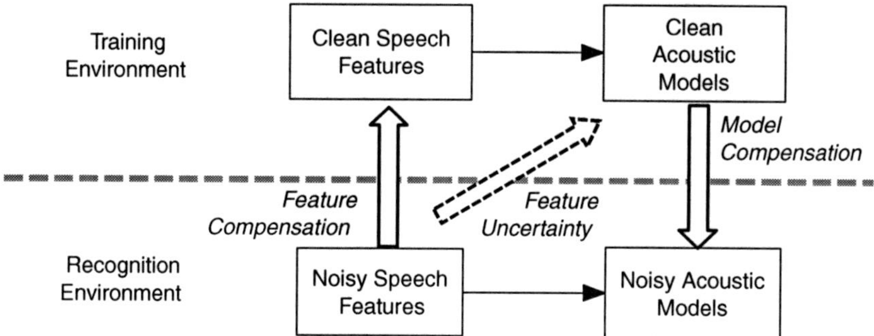

Fig. 6.1 Approaches to noise robust recognition.

models are then mapped so that recognition can take place directly in the noisy environment. In general, feature compensation is simpler and more efficient to implement, but model compensation has the potential for greater robustness since it can make direct use of the very detailed knowledge of the underlying clean speech signal encoded in the acoustic models.

In addition to the above, both feature and model compensation can be augmented by assigning a confidence or uncertainty to each feature. The decoding process is then modified to take account of this additional information. These *uncertainty-based* methods include *missing feature decoding* and *uncertainty decoding*. The following sections discuss each of the above in more detail.

6.1 Feature Enhancement

The basic goal of feature enhancement is to remove the effects of noise from measured features. One of the simplest approaches to this problem is spectral subtraction (SS) [18]. Given an estimate of the noise spectrum in the linear frequency domain n^f, a clean speech spectrum is computed simply by subtracting n^f from the observed spectra. Any desired feature vector, such as mel-cepstral coefficients, can then be derived in the normal way from the compensated spectrum. Although spectral subtraction is widely used, it nevertheless has several problems. Firstly, speech/non-speech detection is needed to isolate segments of the spectrum from which a noise estimate can be made, typically the first second or so of each utterance. This procedure is prone to errors. Secondly, smoothing may lead to an under-estimate of the noise and occasionally errors lead to an over-estimate. Hence, a practical SS scheme takes the form:

$$\hat{x}_i^{\mathsf{f}} = \begin{cases} y_i^{\mathsf{f}} - \alpha n_i^{\mathsf{f}}, & \text{if } \hat{x}_i^{\mathsf{f}} > \beta n_i^{\mathsf{f}} \\ \beta n_i^{\mathsf{f}}, & \text{otherwise,} \end{cases} \tag{6.3}$$

where the f superscript indicates the frequency domain, α is an empirically determined over-subtraction factor and β sets a floor to avoid the clean spectrum going negative [107].

A more robust approach to feature compensation is to compute a minimum mean square error (MMSE) estimate[2]

$$\hat{x}_t = \mathcal{E}\{x_t | y_t\}. \tag{6.4}$$

The distribution of the noisy speech is usually represented as an N-component Gaussian mixture

$$p(y_t) = \sum_{n=1}^{N} c_n \mathcal{N}(y_t; \mu_{\mathsf{y}}^{(n)}, \Sigma_{\mathsf{y}}^{(n)}). \tag{6.5}$$

If x_t and y_t are then assumed to be jointly Gaussian within each mixture component n then

$$\begin{aligned} \mathcal{E}\{x_t | y_t, n\} &= \mu_{\mathsf{x}}^{(n)} + \Sigma_{\mathsf{xy}}^{(n)} (\Sigma_{\mathsf{y}}^{(n)})^{-1}(y_t - \mu_{\mathsf{y}}^{(n)}) \\ &= \mathbf{A}^{(n)} y_t + \mathbf{b}^{(n)} \end{aligned} \tag{6.6}$$

and hence the required MMSE estimate is just a weighted sum of linear predictions

$$\hat{x}_t = \sum_{n=1}^{N} P(n|y_t)(\mathbf{A}^{(n)} y_t + \mathbf{b}^{(n)}), \tag{6.7}$$

where the posterior component probability is given by

$$P(n|y_t) = c_n \mathcal{N}(y_t; \mu_{\mathsf{y}}^{(n)}, \Sigma_{\mathsf{y}}^{(n)})/p(y_t). \tag{6.8}$$

---

[2] The derivation given here considers the complete feature vector y_t to predict the clean speech vector $\hat{x}_t$. It is also possible to just predict the static clean means $\hat{x}_t^{\mathsf{s}}$ and then derive the delta parameters from these estimates.

A number of different algorithms have been proposed for estimating the parameters $\mathbf{A}^{(n)}$, $\mathbf{b}^{(n)}$, $\boldsymbol{\mu}_{\mathsf{y}}^{(n)}$ and $\boldsymbol{\Sigma}_{\mathsf{y}}^{(n)}$ [1, 123]. If so-called *stereo data*, i.e., simultaneous recordings of noisy and clean data is available, then the parameters can be estimated directly. For example, in the SPLICE algorithm [34, 39], the noisy speech GMM parameters $\boldsymbol{\mu}_{\mathsf{y}}^{(n)}$ and $\boldsymbol{\Sigma}_{y}^{(n)}$ are trained directly on the noisy data using EM. For simplicity, the matrix transform $\mathbf{A}^{(n)}$ is assumed to be identity and

$$\mathbf{b}^{(n)} = \frac{\sum_{t=1}^{T} P(n|\boldsymbol{y}_t)(\boldsymbol{x}_t - \boldsymbol{y}_t)}{\sum_{t=1}^{T} P(n|\boldsymbol{y}_t)}. \tag{6.9}$$

A further simplification in the SPLICE algorithm is to find $n^* = \arg\max_n \{P(n|\boldsymbol{y}_t)\}$ and then $\hat{\boldsymbol{x}}_t = \boldsymbol{y}_t + \mathbf{b}^{(n^*)}$.

If stereo data is not available, then the conditional statistics required to estimate $\mathbf{A}^{(n)}$ and $\mathbf{b}^{(n)}$ can be derived using (6.2). One approach to dealing with the non-linearity is to use a first-order vector Taylor series (VTS) expansion [124] to relate the static speech, noise and speech+noise parameters

$$\boldsymbol{y}^{\mathsf{s}} = \boldsymbol{x}^{\mathsf{s}} + f(\boldsymbol{x}_0^{\mathsf{s}}, \boldsymbol{n}_0^{\mathsf{s}}, \boldsymbol{h}_0) + (\boldsymbol{x}^{\mathsf{s}} - \boldsymbol{x}_0^{\mathsf{s}})\frac{\partial f}{\partial \boldsymbol{x}^{\mathsf{s}}}$$
$$+ (\boldsymbol{n}^{\mathsf{s}} - \boldsymbol{n}_0^{\mathsf{s}})\frac{\partial f}{d\boldsymbol{n}^{\mathsf{s}}} + (\boldsymbol{h} - \boldsymbol{h}_0)\frac{\partial f}{\partial \boldsymbol{h}}, \tag{6.10}$$

where the partial derivatives are evaluated at $\boldsymbol{x}_0^{\mathsf{s}}, \boldsymbol{n}_0^{\mathsf{s}}, \boldsymbol{h}_0$. The convolutional term, $\boldsymbol{h}$ is assumed to be a constant convolutional channel effect. Given a clean speech GMM and an initial estimate of the noise parameters, the above function is expanded around the mean vector of each Gaussian and the parameters of the corresponding noisy speech Gaussian are estimated. Then a single iteration of EM is used to update the noise parameters. This ability to estimate the noise parameters without having to explicity identify noise-only segments of the signal is a major advantage of the VTS algorithm.

6.2 Model-Based Compensation

Feature compensation methods are limited by the need to use a very simple model of the speech signal such as a GMM. Given that the acoustic models within the recogniser provide a much more detailed representation of the speech, better performance can be obtained by transforming these models to match the current noise conditions. Model-based compensation is usually based on combining the clean-speech models with a model of the noise and for most situations a single Gaussian noise model is sufficient. However, for some situations with highly varying noise conditions, a multi-state, or multi-Gaussian component, noise model may be useful. This impacts the computational load in both compensation and decoding. If the speech models are M component Gaussians and the noise models are N component Gaussians, then the combined models will have MN components. The same state expansion occurs when combining multiple-state noise models, although in this case, the decoding framework must also be extended to independently track the state alignments in both the speech and noise models [48, 176]. For most purposes, however, single Gaussian noise models are sufficient and in this case the complexity of the resulting models and decoder is unchanged.

The aim of model compensation can be viewed as obtaining the parameters of the speech+noise distribution from the clean speech model and the noise model. Most model compensation methods assume that if the speech and noise models are Gaussian then the combined noisy model will also be Gaussian.

Initially, only the static parameter means and variances will be considered. The parameters of the corrupted speech distribution, $\mathcal{N}(\boldsymbol{\mu}_y, \boldsymbol{\Sigma}_y)$ can be found from

$$\boldsymbol{\mu}_y = \mathcal{E}\left\{\boldsymbol{y}^s\right\} \tag{6.11}$$

$$\boldsymbol{\Sigma}_y = \mathcal{E}\left\{\boldsymbol{y}^s \boldsymbol{y}^{s\mathsf{T}}\right\} - \boldsymbol{\mu}_y \boldsymbol{\mu}_y^{\mathsf{T}}, \tag{6.12}$$

where $\boldsymbol{y}^s$ are the corrupted speech "observations" obtained from a particular speech component combined with noise "observations" from the noise model. There is no simple closed-form solution to the expectation in these equations so various approximations have been proposed.

One approach to model-based compensation is *parallel model combination (PMC)* [57]. Standard PMC assumes that the feature vectors are linear transforms of the log spectrum, such as MFCCs, and that the noise is primarily additive (see [58] for the convolutive noise case).

The basic idea of the algorithm is to map the Gaussian means and variances in the cepstral domain, back into the linear domain where the noise is additive, compute the means and variances of the new combined speech+noise distribution, and then map back to the cepstral domain.

The mapping of a Gaussian $\mathcal{N}(\boldsymbol{\mu}, \boldsymbol{\Sigma})$ from the cepstral to log-spectral domain (indicated by the l superscript) is

$$\boldsymbol{\mu}^{\mathsf{l}} = \mathbf{C}^{-1}\boldsymbol{\mu} \tag{6.13}$$

$$\boldsymbol{\Sigma}^{\mathsf{l}} = \mathbf{C}^{-1}\boldsymbol{\Sigma}(\mathbf{C}^{-1})^{\mathsf{T}}, \tag{6.14}$$

where $\mathbf{C}$ is the DCT. The mapping from the log-spectral to the linear domain (indicated by the f superscript) is

$$\mu_i^{\mathsf{f}} = \exp\{\mu_i^{\mathsf{l}} + \sigma_i^{\mathsf{l}2}/2\} \tag{6.15}$$

$$\sigma_{ij}^{\mathsf{f}} = \mu_i^{\mathsf{f}}\mu_j^{\mathsf{f}}(\exp\{\sigma_{ij}^{\mathsf{l}}\} - 1). \tag{6.16}$$

Combining the speech and noise parameters in this linear domain is now straightforward as the speech and noise are independent

$$\boldsymbol{\mu}_{\mathsf{y}}^{\mathsf{f}} = \boldsymbol{\mu}_{\mathsf{x}}^{\mathsf{f}} + \boldsymbol{\mu}_{\mathsf{n}}^{\mathsf{f}} \tag{6.17}$$

$$\boldsymbol{\Sigma}_{\mathsf{y}}^{\mathsf{f}} = \boldsymbol{\Sigma}_{\mathsf{x}}^{\mathsf{f}} + \boldsymbol{\Sigma}_{\mathsf{n}}^{\mathsf{f}}. \tag{6.18}$$

The distributions in the linear domain will be log-normal and unfortunately the sum of two log-normal distributions is not log-normal. Standard PMC ignores this and assumes that the combined distribution is also log-normal. This is referred to as the *log-normal* approximation. In this case, the above mappings are simply inverted to map $\boldsymbol{\mu}_{\mathsf{y}}^{\mathsf{f}}$ and $\boldsymbol{\Sigma}_{\mathsf{y}}$ back into the cepstral domain. The log-normal approximation is rather poor, especially at low SNRs. A more accurate mapping can be achieved using Monte Carlo techniques to sample the speech and noise distributions, combine the samples using (6.2) and then estimate the new distribution [48]. However, this is computationally infeasible for large systems. A simpler, faster, PMC approximation is to ignore the variances of the speech and noise distributions. This is the *log-add* approximation, here

$$\boldsymbol{\mu}_{\mathsf{y}} = \boldsymbol{\mu}_{\mathsf{x}} + f(\boldsymbol{\mu}_{\mathsf{x}}, \boldsymbol{\mu}_{\mathsf{n}}, \boldsymbol{\mu}_{\mathsf{h}}) \tag{6.19}$$

$$\boldsymbol{\Sigma}_{\mathsf{y}} = \boldsymbol{\Sigma}_{\mathsf{x}}, \tag{6.20}$$

where $f(\cdot)$ is defined in (6.2). This is much faster than standard PMC since the mapping of the covariance matrix to-and-from the cepstral domain is not required.

An alternative approach to model compensation is to use the VTS approximation described earlier in (6.10) [2]. The expansion points for the vector Taylor series are the means of the speech, noise and convolutional noise. Thus for the mean of the corrupted speech distribution

$$\boldsymbol{\mu}_{\mathsf{y}} = \mathcal{E}\left\{\boldsymbol{\mu}_{\mathsf{x}} + f(\boldsymbol{\mu}_{\mathsf{x}}, \boldsymbol{\mu}_{\mathsf{n}}, \boldsymbol{\mu}_{\mathsf{h}}) + (\boldsymbol{x}^{\mathsf{s}} - \boldsymbol{\mu}_{\mathsf{x}})\frac{\partial f}{\partial \boldsymbol{x}^{\mathsf{s}}}\right.$$
$$\left. + (\boldsymbol{n}^{\mathsf{s}} - \boldsymbol{\mu}_{\mathsf{n}})\frac{\partial f}{\partial \boldsymbol{n}^{\mathsf{s}}} + (\boldsymbol{h} - \boldsymbol{\mu}_{\mathsf{h}})\frac{\partial f}{\partial \boldsymbol{h}}\right\}. \tag{6.21}$$

The partial derivatives above can be expressed in terms of partial derivatives of the $\boldsymbol{y}^{\mathsf{s}}$ with respect to $\boldsymbol{x}^{\mathsf{s}}$, $\boldsymbol{n}^{\mathsf{s}}$ and $\boldsymbol{h}$ evaluated at $\boldsymbol{\mu}_{\mathsf{x}}, \boldsymbol{\mu}_{\mathsf{n}}, \boldsymbol{\mu}_{\mathsf{h}}$. These have the form

$$\partial \boldsymbol{y}^{\mathsf{s}}/\partial \boldsymbol{x}^{\mathsf{s}} = \partial \boldsymbol{y}^{\mathsf{s}}/\partial \boldsymbol{h} = \mathbf{A} \tag{6.22}$$

$$\partial \boldsymbol{y}^{\mathsf{s}}/\partial \boldsymbol{n}^{\mathsf{s}} = \mathbf{I} - \mathbf{A}, \tag{6.23}$$

where $\mathbf{A} = \mathbf{CFC}^{-1}$ and $\mathbf{F}$ is a diagonal matrix whose inverse has leading diagonal elements given by $(1 + \exp(\mathbf{C}^{-1}(\boldsymbol{\mu}_{\mathsf{n}} - \boldsymbol{\mu}_{\mathsf{x}} - \boldsymbol{\mu}_{\mathsf{h}})))$. It follows that the means and variances of the noisy speech are given by

$$\boldsymbol{\mu}_{\mathsf{y}} = \boldsymbol{\mu}_{\mathsf{x}} + f(\boldsymbol{\mu}_{\mathsf{x}}, \boldsymbol{\mu}_{\mathsf{n}}, \boldsymbol{\mu}_{\mathsf{h}}) \tag{6.24}$$

$$\boldsymbol{\Sigma}_{\mathsf{y}} = \mathbf{A}\boldsymbol{\Sigma}_{\mathsf{x}}\mathbf{A}^{\mathsf{T}} + \mathbf{A}\boldsymbol{\Sigma}_{\mathsf{h}}\mathbf{A}^{\mathsf{T}} + (\mathbf{I} - \mathbf{A})\boldsymbol{\Sigma}_{\mathsf{n}}(\mathbf{I} - \mathbf{A})^{\mathsf{T}}. \tag{6.25}$$

It can be shown empirically that this VTS approximation is in general more accurate than the PMC log-normal approximation. Note the compensation of the mean is exactly the same as the PMC log-add approximation.

For model-based compensation schemes, an interesting issue is how to compensate the delta and delta–delta parameters [59, 66]. A common approximation is the *continuous time approximation* [66]. Here the dynamic parameters, which are a discrete time estimate of the gradient, are approximated by the instantaneous derivative with respect to time

$$\Delta \boldsymbol{y}_t^{\mathsf{s}} = \frac{\sum_{i=1}^{n} w_i \left(\boldsymbol{y}_{t+i}^{\mathsf{s}} - \boldsymbol{y}_{t-i}^{\mathsf{s}}\right)}{2\sum_{i=1}^{n} w_i^2} \approx \frac{\partial \boldsymbol{y}_t^{\mathsf{s}}}{\partial t}. \tag{6.26}$$

It is then possible to show that, for example, the mean of the delta parameters $\mu_{\Delta y}$ can be expressed as

$$\mu_{\Delta y} = A\mu_{\Delta x}. \tag{6.27}$$

The variances and delta–delta parameters can also be compensated in this fashion.

The compensation schemes described above have assumed that the noise model parameters, μ_n, Σ_n and μ_h, are known. In a similar fashion to estimating the parameters for feature-enhancement ML-estimates of the noise parameters may be found using VTS [101, 123] if this is not the case.

6.3 Uncertainty-Based Approaches

As noted above, whilst feature compensation schemes are computationally efficient, they are limited by the very simple speech models that can be used. On the other hand, model compensation schemes can take full advantage of the very detailed speech model within the recogniser but they are computationally very expensive. The uncertainty-based techniques discussed in this section offer a compromise position.

The effects of additive noise can be represented by a DBN as shown in Figure 6.2. In this case, the state likelihood of a noisy speech feature vector can be written as

$$p(y_t|\theta_t; \lambda, \check{\lambda}) = \int_{x_t} p(y_t|x_t; \check{\lambda})p(x_t|\theta_t; \lambda)dx_t \tag{6.28}$$

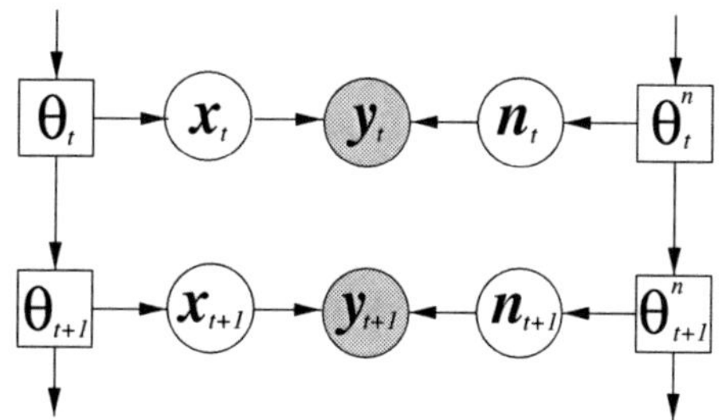

Fig. 6.2 DBN of combined speech and noise model.

where

$$p(\boldsymbol{y}_t|\boldsymbol{x}_t;\check{\boldsymbol{\lambda}}) = \int_{\boldsymbol{n}_t} p(\boldsymbol{y}_t|\boldsymbol{x}_t,\boldsymbol{n}_t)p(\boldsymbol{n}_t;\check{\boldsymbol{\lambda}})d\boldsymbol{n}_t \qquad (6.29)$$

and $\boldsymbol{\lambda}$ denotes the clean speech model and $\check{\boldsymbol{\lambda}}$ denotes the noise model.

The conditional probability in (6.29) can be regarded as a measure of the uncertainty in $\boldsymbol{y}_t$ as a noisy estimate of $\boldsymbol{x}_t$.[3] At the extremes, this could be regarded as a switch such that if the probability is one then $\boldsymbol{x}_t$ is known with certainty otherwise it is completely unknown or *missing*. This leads to a set of techniques called *missing feature techniques*. Alternatively, if $p(\boldsymbol{y}_t|\boldsymbol{x}_t;\check{\boldsymbol{\lambda}})$ can be expressed parametrically, it can be passed directly to the recogniser as a measure of uncertainty in the observed feature vector. This leads to an approach called *uncertainty decoding*.

6.3.1   Missing feature techniques

Missing feature techniques grew out of work on auditory scene analysis motivated by the intuition that humans can recognise speech in noise by identifying strong speech events in the time-frequency plane and ignoring the rest of the signal [30, 31]. In the spectral domain, a sequence of feature vectors $Y_{1:T}$ can be regarded as a spectrogram indexed by $t = 1,\ldots,T$ in the time dimension and by $k = 1,\ldots,K$ in the frequency dimension. Each feature element y_{tk} has associated with it a bit which determines whether or not that "pixel" is reliable. In aggregate, these bits form a noise mask which determine regions of the spectrogram that are hidden by noise and therefore "missing." Speech recognition within this framework therefore involves two problems: finding the mask, and then computing the likelihoods given that mask [105, 143].

The problem of finding the mask is the most difficult aspect of missing feature techniques. A simple approach is to estimate the SNR and then apply a floor. This can be improved by combining it with perceptual criteria such as harmonic peak information, energy ratios, etc. [12]. More robust performance can then be achieved by training a

---

[3] In [6] this uncertainty is made explicit and it is assumed that $p(\boldsymbol{y}_t|\boldsymbol{x}_t;\check{\boldsymbol{\lambda}}) \approx p(\boldsymbol{x}_t|\boldsymbol{y}_t;\check{\boldsymbol{\lambda}})$. This is sometimes called *feature uncertainty*.

Bayesian classifier on clean speech which has been artificially corrupted with noise [155].

Once the noise mask has been determined, likelihoods are calculated either by *feature-vector* imputation or by marginalising out the unreliable features. In the former case, the missing feature values can be determined by a MAP estimate assuming that the vector was drawn from a Gaussian mixture model of the clean speech (see [143] for details and other methods).

Empirical results show that marginalisation gives better performance than imputation. However, when imputation is used, the reconstructed feature vectors can be transformed into the cepstral domain and this outweighs the performance advantage of marginalisation. An alternative approach is to map the spectographic mask directly into cepstral vectors with an associated variance [164]. This then becomes a form of uncertainty decoding discussed next.

6.3.2 Uncertainty Decoding

In uncertainty decoding, the goal is to express the conditional probability in (6.29) in a parametric form which can be passed to the decoder and then used to compute (6.28) efficiently. An important decision is the form that the conditional probability takes. The original forms of uncertainty decoding made the form of the conditional distribution dependent on the region of acoustic space. An alternative approach is to make the form of conditional distribution dependent on the Gaussian component, in a similar fashion to the regression classes in linear transform-based adaptation described in *Adaptation and Normalisation*.

A natural choice for approximating the conditional is to use an N-component GMM to model the acoustic space. Depending on the form of uncertainty being used, this GMM can be used in a number of ways to model the conditional distribution, $p(\boldsymbol{y}_t|\boldsymbol{x}_t;\check{\boldsymbol{\lambda}})$ [38, 102]. For example in [102]

$$p(\boldsymbol{y}_t|\boldsymbol{x}_t;\check{\boldsymbol{\lambda}}) \approx \sum_{n=1}^{N} P(n|\boldsymbol{x}_t;\check{\boldsymbol{\lambda}})p(\boldsymbol{y}_t|\boldsymbol{x}_t,n;\check{\boldsymbol{\lambda}}). \qquad (6.30)$$

If the component posterior is approximated by

$$P(n|\boldsymbol{x}_t;\check{\boldsymbol{\lambda}}) \approx P(n|\boldsymbol{y}_t;\check{\boldsymbol{\lambda}}) \tag{6.31}$$

it is possible to construct a GMM based on the corrupted speech observation, $\boldsymbol{y}_t$, rather than the unseen clean speech observation, $\boldsymbol{x}_t$. If the conditional distribution is further approximated by choosing just the most likely component n^*, then

$$p(\boldsymbol{y}_t|\boldsymbol{x}_t;\check{\boldsymbol{\lambda}}) \approx p(\boldsymbol{y}_t|\boldsymbol{x}_t,\check{c}_{n^*};\check{\boldsymbol{\lambda}}). \tag{6.32}$$

The conditional likelihood in (6.32) has been reduced to a single Gaussian, and since the acoustic model mixture components are also Gaussian, the state/component likelihood in (6.28) becomes the convolution of two Gaussians which is also Gaussian. It can be shown that in this case, the form of (6.28) reduces to

$$p(\boldsymbol{y}_t|\boldsymbol{\theta}_t;\lambda,\check{\boldsymbol{\lambda}}) \propto \sum_{m\in\theta_t} c_m \mathcal{N}(\mathbf{A}^{(n^*)}\boldsymbol{y}_t + \mathbf{b}^{(n^*)};\boldsymbol{\mu}^{(m)},\boldsymbol{\Sigma}^{(m)} + \boldsymbol{\Sigma}_{\mathsf{b}}^{(n^*)}). \tag{6.33}$$

Hence, the input feature vector is linearly transformed in the front-end and at run-time, the only additional cost is that a global bias must be passed to the recogniser and added to the model variances. The same form of compensation scheme is obtained using SPLICE with uncertainty [38], although the way that the compensation parameters, $\mathbf{A}^{(n)}, \mathbf{b}^{(n)}$ and $\boldsymbol{\Sigma}_{\mathsf{b}}^{(n)}$ are computed differs. An interesting advantage of these schemes is that the cost of calculating the compensation parameters is a function of the number of components used to model the feature-space. In contrast for model-compensation techniques such as PMC and VTS, the compensation cost depends on the number of components in the recogniser. This decoupling yields a useful level of flexibility when designing practical systems.

Although uncertainty decoding based on a front-end GMM can yield good performance, in low SNR conditions, it can cause all compensated models to become identical, thus removing all discriminatory power from the recogniser [103]. To overcome this problem, the conditional distribution can alternatively be linked with the recogniser model components. This is similar to using regression classes in adaptation schemes such as MLLR. In this case

$$p(\boldsymbol{y}_t|\theta_t; \boldsymbol{\lambda}, \check{\boldsymbol{\lambda}}) = \sum_{m \in \theta_t} |\mathbf{A}^{(r_m)}| c_m \mathcal{N}(\mathbf{A}^{(r_m)}\boldsymbol{y}_t + \mathbf{b}^{(r_m)}; \boldsymbol{\mu}^{(m)}, \boldsymbol{\Sigma}^{(m)} + \boldsymbol{\Sigma}_{\mathsf{b}}^{(r_m)}),$$

$$(6.34)$$

where r_m is the regression class that component m belongs to. Note that for these schemes, the Jacobian of the transform, $|\mathbf{A}^{(r_m)}|$ can vary from component to component in contrast to the fixed value used in (6.33). The calculation of the compensation parameters follows a similar fashion to the front-end schemes. An interesting aspect of this approach is that as the number of regression classes tends to the number of components in the recogniser, so the performance tends to that of the model-based compensation schemes. Uncertainty decoding of this form can also be incorporated into an adaptive training framework [104].

7

Multi-Pass Recognition Architectures

The previous sections have reviewed some of the basic techniques available for both training and adapting an HMM-based recognition system. In general, any particular combination of model set and adaptation technique will have different characteristics and make different errors. Furthermore, if the outputs of these systems are converted to confusion networks, then it is straightforward to combine the confusion networks and select the word sequence with the overall maximum posterior likelihood. Thus, modern transcription systems typically utilise multiple model sets and make multiple passes over the data. This section starts by describing a typical multi-pass combination framework. This is followed by discussion of how multiple systems may be combined together, as well as how complementary systems may be generated. Finally a description of how these approaches, along with the schemes described earlier, are used for a BN transcription task in three languages: English; Mandarin; and Arabic.

7.1 Example Architecture

A typical architecture is shown in Figure 7.1. A first pass is made over the data using a relatively simple SI model set. The 1-best output

77

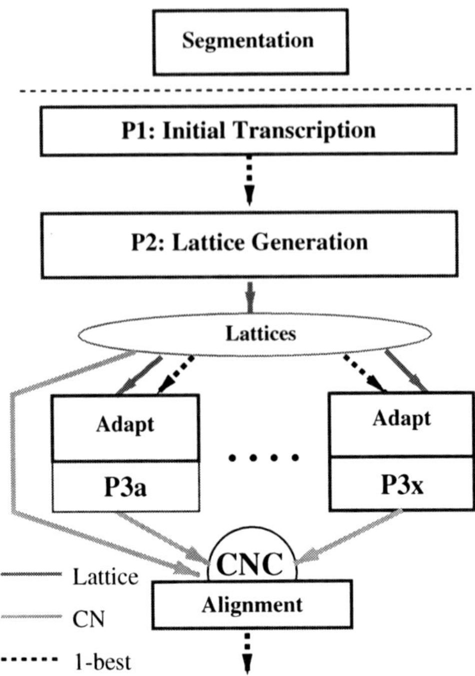

Fig. 7.1 Multi-pass/system combination architectures.

from this pass is used to perform a first round of adaptation. The adapted models are then used to generate lattices using a basic bigram or trigram word-based language model. Once the lattices have been generated, a range of more complex models and adaptation techniques can be applied in parallel to provide K candidate output confusion networks from which the best word sequence is extracted. These 3rd pass models may include ML and MPE trained systems, SI and SAT trained models, triphone and quinphone models, lattice-based MLLR, CMLLR, 4-gram language models interpolated with class-ngrams and many other variants. For examples of recent large-scale transcription systems see [55, 160, 163].

When using these combination frameworks, the form of combination scheme must be determined and the type of individual model set designed. Both these topics will be discussed in the following sections.

7.2 System Combination

Each of the systems, or branches, in Figure 7.1 produces a number of outputs that can be used for combination. Example outputs that can be used are: the hypothesis from each system, $\boldsymbol{w}^{(k)}$ for the kth system; the score from the systems, $p(\mathbf{Y}_{1:T}; \boldsymbol{\lambda}^{(k)})$, as well as the possibility of using the hypotheses and lattices for adaptation and/or rescoring. Log-linear combination of the scores may also be used [15], but this is not common practice in current multi-pass systems.

7.2.1 Consensus Decoding

One approach is to combine either the 1-best hypotheses or confusion network outputs, from the systems together. This may be viewed as an extension of the MBR decoding strategy mentioned earlier. Here

$$\hat{\boldsymbol{w}} = \arg\min_{\tilde{\boldsymbol{w}}} \left\{ \sum_{\boldsymbol{w}} P(\boldsymbol{w}|\boldsymbol{Y}_{1:T}; \boldsymbol{\lambda}^{(1)}, \ldots, \boldsymbol{\lambda}^{(K)}) \mathcal{L}(\boldsymbol{w}, \tilde{\boldsymbol{w}}) \right\}, \qquad (7.1)$$

where $\mathcal{L}(\boldsymbol{w}, \tilde{\boldsymbol{w}})$ is the loss, usually measured in words, between the two word sequences. In contrast to the standard decoding strategy, the word-posterior is a function of multiple systems, $\boldsymbol{\lambda}^{(1)}, \ldots, \boldsymbol{\lambda}^{(K)}$, not a single one (the parameters of this ensemble will be referred to as $\boldsymbol{\lambda}$ in this section).

There are two forms of system combination commonly used: recogniser output voting error reduction (ROVER) [43]; and confusion network combination (CNC) [42]. The difference between the two schemes is that ROVER uses the 1-best system output, whereas CNC uses confusion networks. Both schemes use a two-stage process:

(1) *Align hypotheses/confusion networks*: The output from each of the systems are aligned against one another minimising for example the Levenshtein edit distance (and for CNC the expected edit distance). Figure 7.2 shows the process for combining four system outputs, to form a single word transition network (WTN), similar in structure to a confusion network. For ROVER, confidence scores may also be incorporated into

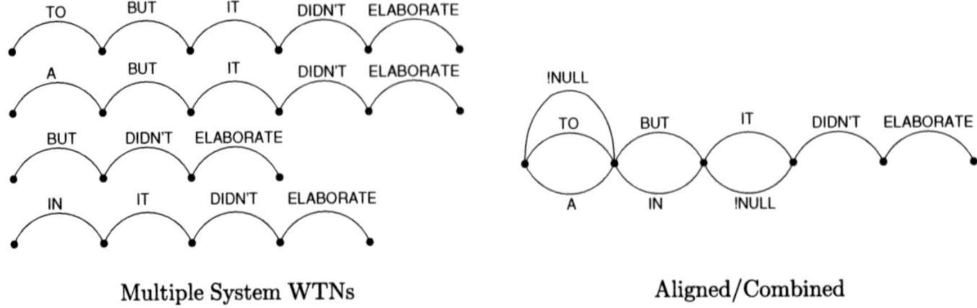

Fig. 7.2 Example of output combination.

the combination scheme (this is essential if only two systems are to be combined).

(2) *Vote/maximum posterior*: Having obtained the WTN (or composite confusion network) decisions need to be made for the final output. As a linear network has been obtained, it is possible to select words for each node-pairing independently. For ROVER the selected word is based on the maximum posterior where the ith word is

$$
P(w_i | \boldsymbol{Y}_{1:T}; \boldsymbol{\lambda})
$$

$$
\approx \frac{1}{K} \sum_{k=1}^{K} \left((1 - \alpha) P(w_i | \boldsymbol{Y}_{1:T}; \boldsymbol{\lambda}^{(k)}) + \alpha \mathcal{D}_i^{(k)}(w_i) \right) \quad (7.2)
$$

$\mathcal{D}_i^{(k)}(w_i) = 1$ when the ith word in the 1-best output is w_i for the kth system and α controls the balance between the 1-best decision and the confidence score.

CNC has been shown to yield small gains over ROVER combination [42].

7.2.2   Implicit Combination

The schemes described above explicitly combine the scores and hypotheses from the various systems. Alternatively, it is possible to implicitly combine the information from one system with another. There are two approaches that are commonly used, lattice or N-best rescoring and cross-adaptation.

Lattice [7, 144] and *N*-best [153] rescoring, as discussed in *Architecture of an HMM-Based Recogniser*, are standard approaches to speeding up decoding by reducing the search-space. Though normally described as approaches to improve efficiency, they have a side-effect of implicit system combination. By restricting the set of possible errors that can be made, the performance of the rescoring system may be improved as it cannot make errors that are not in the *N*-best list or lattices. This is particularly true if the lattices or *N*-best lists are heavily pruned.

Another common implicit combination scheme is cross-adaptation. Here one speech recognition system is used to generate the transcription hypotheses for a second speech recognition system to use for adaptation. The level of information propagated between the two systems is determined by the complexity of the adaptation transforms being used, for example in MLLR the number of linear transforms.

7.3 Complementary System Generation

For the combination schemes above, it is assumed that systems that make different errors are available to combine. Many sites adopt an *ad-hoc* approach to determining the systems to be combined. Initially a number of candidates are constructed. For example: triphone and quinphone models; SAT and GD models; MFCC, PLP, MLP-based posteriors [199] and Gaussianised front-ends; multiple and single pronunciation dictionaries; random decision trees [162]. On a held-out test-set, the performance of the combinations is evaluated and the "best" (possibly taking into account decoding costs) is selected and used. An alternative approach is to train systems that are designed to be complementary to one another.

Boosting is a standard machine learning approach that allows complementary systems to be generated. Initially this was applied to binary classification tasks, for example AdaBoost [44], and later to multi-class tasks, e.g., [45]. The general approach in boosting is to define a distribution, or weight, over the training data, so that correctly classified data is de-weighted and incorrectly classified data is more heavily weighted. A classifier is then trained using this weighted data and a new distribution specified combining the new classifier with all the other classifiers.

There are a number of issues which make the application of boosting, or boosting-like, algorithms to speech recognition difficult. In ASR, the forms of classifier used are highly complex and the data is dynamic. Also, a simple weighted voting scheme is not directly applicable given the potentially vast number of classes in speech recognition. Despite these issues, there have been a number of applications of boosting to speech recognition [37, 121, 154, 201]. An alternative approach, again based on weighting the data, is to modify the form of the decision tree so that leaves are concentrated in regions where the classifier performs poorly [19].

Although initial attempts to explicitly generate complementary systems are promising, the majority of multi-pass systems currently deployed are built using the *ad-hoc* approach described earlier.

7.4 Example Application — Broadcast News Transcription

This section describes a number of systems built at Cambridge University for transcribing BN in three languages, English, Mandarin and Arabic. These illustrate many of the design issues that must be considered when building large-vocabulary multi-pass combination systems. Though the three languages use the same acoustic model structure there are language specific aspects that are exploited to improve performance, or yield models suitable for combination. In addition to BN transcription, the related task of broadcast conversation (BC) transcription is also discussed.

For all the systems described, the front-end processing from the CU-HTK 2003 BN system [87] was used. Each frame of speech is represented by 13 PLP coefficients based on a linear prediction order of 12 with first, second and third-order derivatives appended and then projected down to 39 dimensions using HLDA [92] optimised using the efficient iterative approach described in [51]. Cepstral mean normalisation is applied at the segment level where each segment is a contiguous block of data that has been assigned to the same speaker/acoustic condition.

All models were built using the HTK toolkit [189]. All phones have three emitting state with a left-to-right no-skip topology. Context dependence was provided by using state-clustered cross-word triphone models [190]. In addition to the phone models, two silence models were used: *sil* represents longer silences that break phonetic context; and *sp* represents short silences across which phonetic context is maintained. Gaussian mixture models with an average of 36 Gaussian components per state were used for all systems. The distribution of components over the states was determined based on the state occupancy count [189]. Models were built initially using ML training and then discriminatively trained using MPE training [139, 182]. Gender-dependent versions of these models were then built using discriminative MPE–MAP adaptation [137]. As some BN data, for example telephone interviews, are transmitted over bandwidth-limited channels, wide-band and narrowband spectral analysis variants of each model set were also trained.

The following sections describe how the above general procedure is applied to BN transcription in three different languages; English, Mandarin and Arabic; and BC transcription in Arabic and Mandarin. For each of these tasks the training data, phone-set and any front-end specific processing are described. In addition, examples are given of how a complementary system may be generated.

The language models used for the transcription systems are built from a range of data sources. The first are the transcriptions from the audio data. This will be closely matched to the word-sequences of the test data but is typically limited in size, normally less than 10 million words. In addition, text corpora from news-wire and newspapers are used to increase the quantity of data to around a billion words. Given the nature of the training data, the resulting LMs are expected to be more closely matched to BN-style rather than the more conversational BC-style data.

7.4.1 English BN Transcription

The training data for the English BN transcription system comprised two blocks of data. The first block of about 144 hours had detailed manual transcriptions. The second block of about 1200 hours of data had either closed-caption or similar rough transcriptions provided. This latter data was therefore used in lightly supervised fashion. The dictionary for the English system was based on 45 phones and a 59K word-list was used, which gave out-of-vocabulary (OOV) rates of less than 1%. Two dictionaries were constructed. The first was a multiple pronunciation dictionary (MPron), having about 1.1 pronunciations per word on average. The second was a single pronunciation dictionary (SPron) built using the approach described in [70]. This is an example of using a modified dictionary to build complementary systems. HMMs were built using each of these dictionaries with about 9000 distinct states. A four-gram language model was used for all the results quoted here. This was trained using 1.4 billion words of data, which included the acoustic training data references and closed-captions.

In order to assess the performance of the English BN transcription system, a range of development and evaluation data sets were used. Results are given here on three separate test sets. dev04 and dev04f were each 3 hours in size. dev04f was expected to be harder to recognise as it contains more challenging data with high levels of background noise/music and non-native speakers. eval04 was a 6 hour test set evenly split between data similar to dev04 and dev04f.

Table 7.1 shows the performance of the MPron and the SPron systems using the architecture shown in Figure 7.1. Here a segmentation and clustering supplied by the Computer Sciences Laboratory for Mechanics and Engineering Sciences (LIMSI), a French CNRS laboratory, was used in the decoding experiments. There are a number of points illustrated by these results. First, the P3c branch is actually an example of cross-adaptation since the MPron hypotheses from the P2-stage are used in the initial transform estimation for the SPron system. By themselves the SPron and MPron systems perform about

Table 7.1 %WER of using MPron and SPron acoustic models and the LIMSI segmentation/clustering for English BN transcription.

System		%WER		
		dev04	dev04f	eval04
P2	MPron	11.1	15.9	13.6
P3b	MPron	10.6	15.4	13.4
P3c	SPron	10.4	15.2	13.0
P2+P3b	CNC	10.8	15.3	13.3
P2+P3c	CNC	10.3	14.8	12.8
P2+P3b+P3c	CNC	10.5	15.1	12.9

the same [56], but the P3c branch is consistently better than the P3b branch. This is due to cross-adaptation. Combining the P3c branch with the P2 branch using CNC (P2+P3c) shows further gains of 0.2%–0.4%. This shows the advantage of combining systems with different acoustic models. Finally, combining the P2, P3b and P3c branches together gave no gains, indeed slight degradations compared to the P2+P3c combination can be seen. This illustrates one of the issues with combining systems together. The confidence scores used are not accurate representations of whether the word is correct or not, thus combining systems may degrade performance. This is especially true when the systems have very different performance, or two of the systems being combined are similar to one another.

It is interesting to note that the average performance over the various test sets varies from 10.5% (dev04) to 15.1% (dev04f). This reflects the more challenging nature of the dev04f data. Thus, even though only test data classified as BN is being used, the error rates vary by about 50% between the test sets.

In addition to using multiple forms of acoustic model, it is also possible to use multiple forms of segmentation [56]. The first stage in processing the audio is to segment the audio data into homogeneous blocks.[1] These blocks are then clustered together so that all the data from the speaker/environment are in the same cluster. This process, sometimes referred to as *diarisation*, may be scored as a task in itself [170]. However, the interest here is how this segmentation and clustering affects the system combination performance. CNC could be used to combine the output from the two segmentations but here ROVER was used. To perform segmentation combination, separate branches from the initial P1–P2 stage are required. This architecture is shown in Figure 7.3.

The results using this dual architecture are shown in Table 7.2. The two segmentations used were: the LIMSI segmentation and clustering used to generate the results in Table 7.1; and the segmentation generated at Cambridge University Engineering Department (CUED)

---

[1] Ideally, contiguous blocks of data that have the same speaker and the same acoustic/ channel conditions.

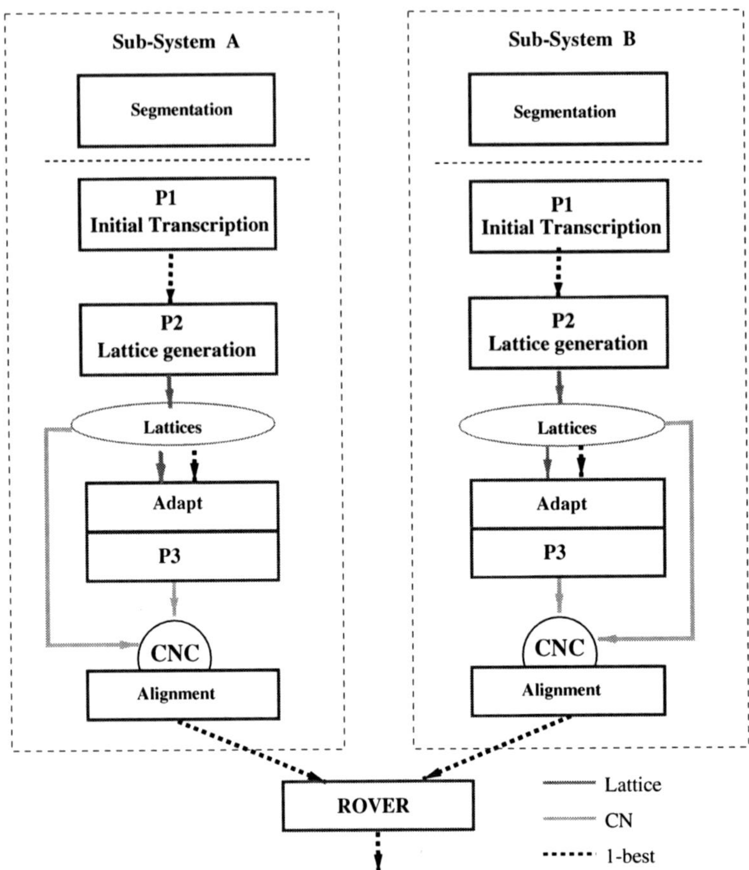

Fig. 7.3 Multi-pass/system cross segmentation combination architecture.

Table 7.2 %WER of the dual segmentation system for English BN transcription).

Segmentation/	%WER		
Clustering	dev04	dev04f	eval04
LIMSI	10.3	14.9	12.8
CUED	10.4	15.2	12.9
LIMSI $\oplus$ CUED	9.8	14.7	12.4

using different algorithms [161]. Thus, the systems differ in both the segmentation and clustering fed into the recognition system [56].

For both segmentation and clusterings, the equivalent of the P2+P3c branch in Table 7.1 were used. Although the segmentations have similar average lengths 13.6 and 14.2 seconds for the LIMSI and CUED, respectively, they have different numbers of clusters and diarisation scores. The slight difference in performance for the LIMSI segmentations was because slightly tighter beams for P2 and P3c branches were used to keep the overall run-time approximately the same. Combining the two segmentations together gave gains of between 0.2% and 0.5%.

7.4.2   Mandarin BN and BC Transcription

About 1390 hours of acoustic training data was used to build the Mandarin BN transcription system. In contrast to the English BN system, the Mandarin system was required to recognise both BN and BC data. The training data was approximately evenly split between these two types. For this task, manual transcriptions were available for all the training data. As Mandarin is a tonal language, fundamental frequency (F0) was added to the feature vector after the application of the HLDA transform. Delta and delta–delta F0 features were also added to give a 42 dimensional feature vector. The phone-set used for these systems was derived from the 60 toneless phones used in the 44K dictionary supplied by the Linguistic Data Consortium (LDC). This set is the same as the one used in CU-HTK 2004 Mandarin CTS system [160]. It consists of 46 toneless phones, obtained from the 60 phone LDC set by applying mappings of the form "[aeiu]n→[aeiu] n", "[aeiou]ng→[aeiou] ng" and "u:e→ue". In addition to the F0 features, the influence of tone on the acoustic models can be incorporated into the acoustic model units. Most dictionaries for Mandarin give tone-markers for the vowels. This tone information may be used either by treating each tonal vowel as a separate acoustic unit, or by including tone questions about both the centre and surround phones into the decision trees used for phonetic context clustering. The latter approach is adopted in these experiments.

One issue in language modelling for the Chinese language is that there are no natural word boundaries in normal texts. A string of characters may be partitioned into "words" in a range of ways and there are multiple valid partitions that may be used. As in the CU-HTK Mandarin CTS system [160], the LDC character to word segmenter was used. This implements a simple deepest first search method to determine the word boundaries. Any Chinese characters that are not present in the word list are treated as individual words. The Mandarin word-list required for this part of the process was based on the 44K LDC Mandarin dictionary. As some English words were present in the acoustic training data transcriptions, all English words and single character Mandarin words not in the LDC list were added to the word-list to yield a total of about 58K words. Note that the Mandarin OOV rate is therefore effectively zero since pronunciations were available for all single characters. A four-gram language model was constructed at the word level using 1.3 billion "words" of data.

The performance of the Mandarin system was evaluated on three test sets. `bnmdev06` consists of 3 hours of BN-style data, whereas `bcmdev05` comprises about 3 hours of BC-style data. `eval06` is a mixture of the two styles. As word boundaries are not usually given in Mandarin, performance is assessed in terms of the *character error rate* (CER). For this task the average number of characters per word is about 1.6.

A slightly modified version of the architecture shown in Figure 7.1 was used for these experiments. The P2 output was not combined with the P3 branches to produce the final output, instead two P3 branches in addition to the HLDA system (P3b) were generated. A SAT system using CMLLR transforms to represent each speaker (HLDA-SAT) was used for branch P3a. Gaussianisation was applied after the HLDA transform, using the GMM approach with 4 components for each dimension, to give a GAUSS system for branch P3c. The same adaptation stages were used for all P3 branches irrespective of the form of model.

Table 7.3 shows the performance of the various stages. For all systems, the performance on the BC data is, as expected, worse than the performance on the BN data. The BC data is more conversational in style and thus inherently harder to recognise as well as not being as well matched to the language model training data. The best performing single branch was the GAUSS branch. For this task the performance of the HLDA-SAT system was disappointing, giving only small gains if

Table 7.3 %CER using a multi-pass combination scheme for Mandarin BN and BC transcription.

System		%CER		
		bnmdev06	bcmdev05	eval06
P2	HLDA	8.3	18.8	18.7
P3a	HLDA-SAT	8.0	18.1	17.9
P3b	HLDA	7.9	18.2	18.1
P3c	GAUSS	7.9	17.7	17.4
P3a+P3c	CNC	7.7	17.6	17.3

any. On a range of tasks, SAT training, though theoretically interesting, has not been found to yield large gains in current configurations. In common with the English system, however, small consistent gains were obtained by combining multiple branches together.

7.4.3 Arabic BN and BC Transcription

Two forms of Arabic system are commonly constructed: graphemic and phonetic systems. In a graphemic system, a dictionary is generated using one-to-one letter-to-sound rules for each word. Note that normally the Arabic text is *romanised* and the word and letter order swapped to be left-to-right. Thus, the dictionary entry for the word "book" in Arabic is

```
ktAb          /k/ /t/ /A/ /b/.
```

This scheme yields 28 consonants, four *alif* variants (*madda* and *hamza* above and below), *ya* and *wa* variants (*hamza* above), *ta-marbuta* and *hamza*. Thus, the total number of graphemes is 36 (excluding silence). This simple graphemic representation allows a pronunciation to be automatically generated for any word in the dictionary. In Arabic, the short vowels (*fatha* /a/, *kasra* /i/ and *damma* /u/) and diacritics (*shadda*, *sukun*) are commonly not marked in texts. Additionally, *nunation* can result in a word-final *nun* (/n/) being added to nouns and adjectives in order to indicate that they are unmarked for definiteness. In graphemic systems the acoustic models are required to model the implied pronunciation variations implicitly. An alternative approach is to use a phonetic system where the pronunciations for each word explicitly include the short vowels and *nun*. In this case, for example, the lexicon entry for the word book becomes

`ktAb`   /k/ /i/ /t/ /A/ /b/.

A common approach to obtaining these phonetic pronunciations is to use the Buckwalter Morphological Analyser (version 2.0)[2] [3]. Buckwalter yields pronunciations for about 75% of the words in the 350 K dictionary used in the experiments. This means that pronunciations must be added by hand, or rules used to derive them, or the word removed from the word-list. In this work, only a small number of words were added by hand. In order to reduce the effect of inconsistent diacritic markings (and to make the phonetic system differ from the graphemic system) the variants of *alef*, *ya* and *wa* were mapped to their simple forms. This gives a total of 32 "phones" excluding silence.

With this phonetic system there are an average of about 4.3 pronunciations per word, compared to English with about 1.1 pronunciations. This means that the use of pronunciation probabilities is important. Two word-lists were used: a 350 K word-list for the graphemic system, derived using word frequency counts only; and a 260 K subset of the graphemic word-list comprising words for which phonetic pronunciations could be found. Thus two language models were built, the first using the 350 K graphemic word-list, LM1, and the second using the 260 K phonetic word-list. About 1 billion words of language model training data were used and the acoustic models were trained on about 1000 hours of acoustic data, using the same features as for the English system.

An additional complication is that Arabic is a highly inflected agglutinative language. This results in a potentially very large vocabulary, as words are often formed by attaching affixes to triconsonantal roots, which may result in a large OOV rate compared to an equivalent English system for example. Using a 350 K vocabulary, the OOV rate on `bcat06` was 2.0%, higher than the OOV rate in English with a 59 K vocabulary.

---

[2] Available at http://www.qamus.org/index.html.

Table 7.4 %WER using a multi-pass combination scheme for Arabic BN and BC transcription.

System		WER (%)		
		bnat06	bcat06	eval06
P2a-LM1	Graphemic	21.1	27.6	23.9
P2b-LM2	Graphemic	21.3	27.8	23.8
P3a-LM1	Graphemic	20.4	27.2	23.3
P3b-LM2	Phonetic	20.0	27.3	22.4
P3a+P3b	CNC	18.7	25.3	21.2

Three test sets defined by BBN Technologies (referred to as BBN) were used for evaluating the systems. The first, bnat06, consists of about 3 hours of BN-style data. The second, bcat06 consists of about 3 hours of BC-style data. The eval06 is approximately 3 hours of a mixture of the two styles.

Table 7.4 shows the performance of the Arabic system using the multi-pass architecture in Figure 7.3. However, rather than using two different segmentations, in this case graphemic and phonetic branches were used. Given the large number of pronunciations and the word-list size, the lattices needed for the P3 rescoring were generated using the graphemic acoustic models for both branches. Thus, the phonetic system has an element of cross-adaptation between the graphemic and the phonetic systems. Even with the cross-adaptation gains, the graphemic system is still slightly better than the phonetic system on the BC data. However, the most obvious effect is that performance gain obtained from combining the phonetic and graphemic systems using CNC is large, around 6% relative reduction over the best individual branch.

7.4.4   Summary

The above experiments have illustrated a number of points about large vocabulary continuous speech recognition.

In terms of the underlying speech recognition technology, the same broad set of techniques can be used for multiple languages, in this case English, Arabic and Mandarin. However, to enable the technology to work well for different languages, modifications beyond specifying the phone-units and dictionary are required. For example in Mandarin, character-to-word segmentation must be considered, as well as the inclusion of tonal features.

The systems described above have all been based on HLDA feature projection/decorrelation. This, along with schemes such as global STC, are standard approaches used by many of the research groups developing large vocabulary transcription systems. Furthermore, all the acoustic models described have been trained using discriminative, MPE, training. These discriminative approaches are also becoming standard since typical gains of around 10% relative compared to ML can be obtained [56].

Though no explicit contrasts have been given, all the systems described make use of a combination of feature normalisation and linear adaptation schemes. For large vocabulary speech recognition tasks where the test data may comprise a range of speaking styles and accents, these techniques are essential.

As the results have shown, significant further gains can be obtained by combining multiple systems together. For example in Arabic, by combining graphemic and phonetic systems together, gains of 6% relative over the best individual branch were obtained. All the systems described here were built at CUED. Interestingly, even greater gains can be obtained by using *cross-site* system combination [56].

Though BN and BC transcription is a good task to illustrate many of the schemes discussed in this review, rather different tasks would be needed to properly illustrate the various noise robustness techniques described and space prevented inclusion of these. Some model-based schemes have been applied to transcription tasks [104], however, the gains have been small since the levels of background noise encountered in this type of transcription are typically low.

The experiments described have also avoided issues of computational efficiency. For practical systems this is an important consideration, but beyond the scope of this review.

Conclusions

This review has reviewed the core architecture of an HMM-based CSR system and outlined the major areas of refinement incorporated into modern-day systems including those designed to meet the demanding task of large vocabulary transcription.

HMM-based speech recognition systems depend on a number of assumptions: that speech signals can be adequately represented by a sequence of spectrally derived feature vectors; that this sequence can be modelled using an HMM; that the feature distributions can be modelled exactly; that there is sufficient training data; and that the training and test conditions are matched. In practice, all of these assumptions must be relaxed to some extent and it is the extent to which the resulting approximations can be minimised which determines the eventual performance.

Starting from a base of simple single-Gaussian diagonal covariance continuous density HMMs, a range of refinements have been described including distribution and covariance modelling, discriminative parameter estimation, and algorithms for adaptation and noise compensation. All of these aim to reduce the inherent approximation in the basic HMM framework and hence improve performance.

In many cases, a number of alternative approaches have been described amongst which there was often no clear winner. Usually, the various options involve a trade-off between factors such as the availability of training or adaptation data, run-time computation, system complexity and target application. The consequence of this is that building a successful large vocabulary transcription system is not just about finding elegant solutions to statistical modelling and classification problems, it is also a challenging system integration problem. As illustrated by the final presentation of multi-pass architectures and actual example configurations, the most challenging problems require complex solutions.

HMM-based speech recognition is a maturing technology and as evidenced by the rapidly increasing commercial deployment, performance has already reached a level which can support viable applications. Progress is continuous and error rates continue to fall. For example, error rates on the transcription of conversational telephone speech were around 50% in 1995. Today, with the benefit of more data, and the refinements described in this review, error rates are now well below 20%.

Nevertheless, there is still much to do. As shown by the example performance figures given for transcription of broadcast news and conversations, results are impressive but still fall short of human capability. Furthermore, even the best systems are vulnerable to spontaneous speaking styles, non-native or highly accented speech and high ambient noise. Fortunately, as indicated in this review, there are many ideas which have yet to be fully explored and many more still to be conceived.

Finally, it should be acknowledged that despite their dominance and the continued rate of improvement, many argue that the HMM architecture is fundamentally flawed and that performance must asymptote at some point. This is undeniably true, however, no good alternative to the HMM has been found yet. In the meantime, the authors of this review believe that the performance asymptote for HMM-based speech recognition is still some way away.

Acknowledgments

The authors would like to thank all research students and staff, past and present, who have contributed to the development of the HTK toolkit and HTK-derived speech recognition systems and algorithms. In particular, the contribution of Phil Woodland to all aspects of this research and development is gratefully acknowledged. Without all of these people, this review would not have been possible.

Notations and Acronyms

The following notation has been used throughout this review.

$\boldsymbol{y}_t$	speech observation at time t
$\boldsymbol{Y}_{1:T}$	observation sequence $\boldsymbol{y}_1, \ldots, \boldsymbol{y}_T$
$\boldsymbol{x}_t$	"clean" speech observation at time t
$\boldsymbol{n}_t$	"noise" observation at time t
$\boldsymbol{y}_t^{\mathsf{s}}$	the "static" elements of feature vector $\boldsymbol{y}_t$
$\Delta \boldsymbol{y}_t^{\mathsf{s}}$	the "delta" elements of feature vector $\boldsymbol{y}_t$
$\Delta^2 \boldsymbol{y}_t^{\mathsf{s}}$	the "delta–delta" elements of feature vector $\boldsymbol{y}_t$
$\boldsymbol{Y}^{(r)}$	rth observation sequence $\boldsymbol{y}_1^{(r)}, \ldots, \boldsymbol{y}_{T^{(r)}}^{(r)}$
c_m	the prior of Gaussian component m
$\boldsymbol{\mu}^{(m)}$	the mean of Gaussian component m
$\mu_i^{(m)}$	element i of mean vector $\boldsymbol{\mu}^{(m)}$
$\boldsymbol{\Sigma}^{(m)}$	the covariance matrix of Gaussian component m
$\sigma_i^{(m)2}$	ith leading diagonal element of covariance matrix $\boldsymbol{\Sigma}^{(m)}$
$\mathcal{N}(\boldsymbol{\mu}, \boldsymbol{\Sigma})$	Gaussian distribution with mean $\boldsymbol{\mu}$ and covariance matrix $\boldsymbol{\Sigma}$

$\mathcal{N}(\boldsymbol{y}_t; \boldsymbol{\mu}, \boldsymbol{\Sigma})$	likelihood of observation $\boldsymbol{y}_t$ being generated by $\mathcal{N}(\boldsymbol{\mu}, \boldsymbol{\Sigma})$
$\mathbf{A}$	transformation matrix
$\mathbf{A}^{\mathsf{T}}$	transpose of $\mathbf{A}$
$\mathbf{A}^{-1}$	inverse of $\mathbf{A}$
$\mathbf{A}_{[p]}$	top p rows of $\mathbf{A}$
$|\mathbf{A}|$	determinant of $\mathbf{A}$
a_{ij}	element i, j of $\mathbf{A}$
$\mathbf{b}$	bias column vector
$\boldsymbol{\lambda}$	set of HMM parameters
$p(\boldsymbol{Y}_{1:T}; \boldsymbol{\lambda})$	likelihood of observation sequence $\boldsymbol{Y}_{1:T}$ given model parameters $\boldsymbol{\lambda}$
θ_t	state/component occupied at time t
$\mathbf{s}_j$	state j of the HMM
$\mathbf{s}_{jm}$	component m of state j of the HMM
$\gamma_t^{(j)}$	$P(\theta_t = \mathbf{s}_j | \boldsymbol{Y}_{1:T}; \boldsymbol{\lambda})$
$\gamma_t^{(jm)}$	$P(\theta_t = \mathbf{s}_{jm} | \boldsymbol{Y}_{1:T}; \boldsymbol{\lambda})$
$\alpha_t^{(j)}$	forward likelihood, $p(\boldsymbol{y}_1, \ldots, \boldsymbol{y}_t, \theta_t = \mathbf{s}_j; \boldsymbol{\lambda})$
$\beta_t^{(j)}$	backward likelihood, $p(\boldsymbol{y}_{t+1}, \ldots, \boldsymbol{y}_T | \theta_t = \mathbf{s}_j; \boldsymbol{\lambda})$
$d_j(t)$	probability of occupying state $\mathbf{s}_j$ for t time steps
q	a phone such as /a/, /t/, etc
$\boldsymbol{q}^{(w)}$	a pronunciation for word w, $q_1, \ldots, q_{K_w}$
$P(\boldsymbol{w}_{1:L})$	probability of word sequence $\boldsymbol{w}_{1:L}$
$\boldsymbol{w}_{1:L}$	word sequence, $w_1, \ldots, w_L$

Acronyms

BN	broadcast news
BC	broadcast conversations
CAT	cluster adaptive training
CDF	cumulative density function
CMN	cepstral mean normalisation
CML	conditional maximum likelihood
CMLLR	constrained maximum likelihood linear regression
CNC	confusion network combination

CTS	conversation transcription system
CVN	cepstral variance normalisation
DBN	dynamic Bayesian network
EBW	extended Baum-Welch
EM	expectation maximisation
EMLLT	extended maximum likelihood linear transform
FFT	fast Fourier transform
GAUSS	Gaussianisation
GD	gender dependent
GMM	Gaussian mixture model
HDA	heteroscedastic discriminant analysis
HLDA	heteroscedastic linear discriminant analysis
HMM	hidden Markov model
HTK	hidden Markov model toolkit (see http://htk.eng.cam.ac.uk)
LDA	linear discriminant analysis
LM	language model
MAP	maximum *a posteriori*
MBR	minimum Bayes risk
MCE	minimum classification error
MFCC	mel-frequency cepstral coefficients
ML	maximum likelihood
MLLR	maximum likelihood linear regression
MLLT	maximum likelihood linear transform
MLP	multi-layer perceptron
MMI	maximum mutual information
MMSE	minimum mean square error
MPE	minimum phone error
PLP	perceptual linear prediction
PMC	parallel model combination
RSW	reference speaker weighting
PMF	probability mass function
ROVER	recogniser output voting error reduction
SAT	speaker adaptive training
SI	speaker independent

SNR	signal to noise ratio
SPAM	sub-space constrained precision and means
VTLN	vocal tract length normalisation
VTS	vector Taylor series
WER	word error rate
WTN	word transition network

References

[1] A. Acero, *Acoustical and Environmental Robustness in Automatic Speech Recognition.* Kluwer Academic Publishers, 1993.

[2] A. Acero, L. Deng, T. Kristjansson, and J. Zhang, "HMM adaptation using vector Taylor series for noisy speech recognition," in *Proceedings of ICSLP*, Beijing, China, 2000.

[3] M. Afify, L. Nguyen, B. Xiang, S. Abdou, and J. Makhoul, "Recent progress in Arabic broadcast news transcription at BBN," in *Proceedings of Interspeech*, Lisbon, Portugal, September 2005.

[4] S. M. Ahadi and P. C. Woodland, "Combined Bayesian and predictive techniques for rapid speaker adaptation of continuous density hidden Markov models," *Computer Speech and Language*, vol. 11, no. 3, pp. 187–206, 1997.

[5] T. Anastasakos, J. McDonough, R. Schwartz, and J. Makhoul, "A compact model for speaker adaptive training," in *Proceedings of ICSLP*, Philadelphia, 1996.

[6] J. A. Arrowood, *Using Observation Uncertainty for Robust Speech Recognition.* PhD thesis, Georgia Institute of Technology, 2003.

[7] X. Aubert and H. Ney, "Large vocabulary continuous speech recognition using word graphs," in *Proceedings of ICASSP*, vol. 1, pp. 49–52, Detroit, 1995.

[8] S. Axelrod, R. Gopinath, and P. Olsen, "Modelling with a subspace constraint on inverse covariance matrices," in *Proceedings of ICSLP*, Denver, CO, 2002.

[9] L. R. Bahl, P. F. Brown, P. V. de Souza, and R. L. Mercer, "Maximum mutual information estimation of hidden Markov model parameters for speech recognition," in *Proceedings of ICASSP*, pp. 49–52, Tokyo, 1986.

[10] L. R. Bahl, P. V. de Souza, P. S. Gopalakrishnan, D. Nahamoo, and M. A. Picheny, "Robust methods for using context-dependent features and models

in a continuous speech recognizer," in *Proceedings of ICASSP*, pp. 533–536, Adelaide, 1994.

[11] J. K. Baker, "The Dragon system — An overview," *IEEE Transactions on Acoustics Speech and Signal Processing*, vol. 23, no. 1, pp. 24–29, 1975.

[12] J. Barker, M. Cooke, and P. D. Green, "Robust ASR based on clean speech models: an evaluation of missing data techniques for connected digit recognition in noise," in *Proceedings of Eurospeech*, pp. 213–216, Aalberg, Denmark, 2001.

[13] L. E. Baum and J. A. Eagon, "An inequality with applications to statistical estimation for probabilistic functions of Markov processes and to a model for ecology," *Bulletin of American Mathematical Society*, vol. 73, pp. 360–363, 1967.

[14] L. E. Baum, T. Petrie, G. Soules, and N. Weiss, "A maximisation technique occurring in the statistical analysis of probabilistic functions of Markov chains," *Annals of Mathematical Statistics*, vol. 41, pp. 164–171, 1970.

[15] P. Beyerlein, "Discriminative model combination," in *Proceedings of ASRU*, Santa Barbara, 1997.

[16] J. A. Bilmes, "Buried Markov models: A graphical-modelling approach to automatic speech recognition," *Computer Speech and Language*, vol. 17, no. 2–3, 2003.

[17] J. A. Bilmes, "Graphical models and automatic speech recognition," in *Mathematical Foundations of Speech and Language: Processing Institute of Mathematical Analysis Volumes in Mathematics Series*, Springer-Verlag, 2003.

[18] S. Boll, "Suppression of acoustic noise in speech using spectral subtraction," *IEEE Transactions on Acoustics Speech and Signal Processing*, vol. 27, pp. 113–120, 1979.

[19] C. Breslin and M. J. F. Gales, "Complementary system generation using directed decision trees," in *Proceedings of ICSLP*, Toulouse, 2006.

[20] P. F. Brown, *The Acoustic-Modelling Problem in Automatic Speech Recognition*. PhD thesis, Carnegie Mellon, 1987.

[21] P. F. Brown, V. J. Della Pietra, P. V. de Souza, J. C. Lai, and R. L. Mercer, "Class-based N-gram models of natural language," *Computational Linguistics*, vol. 18, no. 4, pp. 467–479, 1992.

[22] W. Byrne, "Minimum Bayes risk estimation and decoding in large vocabulary continuous speech recognition," *IEICE Transactions on Information and Systems: Special Issue on Statistical Modelling for Speech Recognition*, vol. E89-D(3), pp. 900–907, 2006.

[23] H. Y. Chan and P. C. Woodland, "Improving broadcast news transcription by lightly supervised discriminative training," in *Proceedings of ICASSP*, Montreal, Canada, March 2004.

[24] K. T. Chen, W. W. Liao, H. M. Wang, and L. S. Lee, "Fast speaker adaptation using eigenspace-based maximum likelihood linear regression," in *Proceedings of ICSLP*, Beijing, China, 2000.

[25] S. F. Chen and J. Goodman, "An empirical study of smoothing techniques for language modelling," *Computer Speech and Language*, vol. 13, pp. 359–394, 1999.

[26] S. S. Chen and R. Gopinath, "Gaussianization," in *NIPS 2000*, Denver, CO, 2000.

[27] S. S. Chen and R. A. Gopinath, "Model selection in acoustic modelling," in *Proceedings of Eurospeech*, pp. 1087–1090, Rhodes, Greece, 1997.

[28] W. Chou, "Maximum *a-posterior* linear regression with elliptical symmetric matrix variants," in *Proceedings of ICASSP*, pp. 1–4, Phoenix, USA, 1999.

[29] W. Chou, C. H. Lee, and B. H. Juang, "Minimum error rate training based on N-best string models," in *Proceedings of ICASSP*, pp. 652–655, Minneapolis, 1993.

[30] M. Cooke, P. D. Green, and M. D. Crawford, "Handling missing data in speech recognition," in *Proceedings of ICSLP*, pp. 1555–1558, Yokohama, Japan, 1994.

[31] M. Cooke, A. Morris, and P. D. Green, "Missing data techniques for robust speech recognition," in *Proceedings of ICASSP*, pp. 863–866, Munich, Germany, 1997.

[32] S. B. Davis and P. Mermelstein, "Comparison of parametric representations for monosyllabic word recognition in continuously spoken sentences," *IEEE Transactions on Acoustics Speech and Signal Processing*, vol. 28, no. 4, pp. 357–366, 1980.

[33] A. P. Dempster, N. M. Laird, and D. B. Rubin, "Maximum likelihood from incomplete data via the EM algorithm," *Journal of the Royal Statistical Society Series B*, vol. 39, pp. 1–38, 1977.

[34] L. Deng, A. Acero, M. Plumpe, and X. D. Huang, "Large-vocabulary speech recognition under adverse acoustic environments," in *Proceedings of ICSLP*, pp. 806–809, Beijing, China, 2000.

[35] V. Diakoloukas and V. Digalakis, "Maximum likelihood stochastic transformation adaptation of hidden Markov models," *IEEE Transactions on Speech and Audio Processing*, vol. 7, no. 2, pp. 177–187, 1999.

[36] V. Digalakis, H. Collier, S. Berkowitz, A. Corduneanu, E. Bocchieri, A. Kannan, C. Boulis, S. Khudanpur, W. Byrne, and A. Sankar, "Rapid speech recognizer adaptation to new speakers," in *Proceedings of ICASSP*, pp. 765–768, Phoenix, USA, 1999.

[37] C. Dimitrakakis and S. Bengio, "Boosting HMMs with an application to speech recognition," in *Proceedings of ICASSP*, Montreal, Canada, 2004.

[38] J. Droppo, A. Acero, and L. Deng, "Uncertainty decoding with SPLICE for noise robust speech recognition," in *Proceedings of ICASSP*, Orlando, FL, 2002.

[39] J. Droppo, L. Deng, and A. Acero, "Evaluation of the SPLICE algorithm on the aurora 2 database," in *Proceedings of Eurospeech*, pp. 217–220, Aalberg, Denmark, 2001.

[40] G. Evermann, H. Y. Chan, M. J. F. Gales, T. Hain, X. Liu, D. Mrva, L. Wang, and P. Woodland, "Development of the 2003 CU-HTK conversational telephone speech transcription system," in *Proceedings of ICASSP*, Montreal, Canada, 2004.

[41] G. Evermann and P. C. Woodland, "Large vocabulary decoding and confidence estimation using word posterior probabilities," in *Proceedings of ICASSP*, pp. 1655–1658, Istanbul, Turkey, 2000.

[42] G. Evermann and P. C. Woodland, "Posterior probability decoding, confidence estimation and system combination," in *Proceedings of Speech Transcription Workshop*, Baltimore, 2000.

[43] J. Fiscus, "A post-processing system to yield reduced word error rates: Recogniser output voting error reduction (ROVER)," in *Proceedings of IEEE ASRU Workshop*, pp. 347–352, Santa Barbara, 1997.

[44] Y. Freund and R. Schapire, "Experiments with a new boosting algorithm," *Proceedings of the Thirteenth International Conference on Machine Learning*, 1996.

[45] Y. Freund and R. Schapire, "A decision theoretic generalization of on-line learning and an application to boosting," *Journal of Computer and System Sciences*, pp. 119–139, 1997.

[46] K. Fukunaga, *Introduction to Statistical Pattern Recognition*. Academic Press, 1972.

[47] S. Furui, "Speaker independent isolated word recognition using dynamic features of speech spectrum," *IEEE Transactions ASSP*, vol. 34, pp. 52–59, 1986.

[48] M. J. F. Gales, *Model-based Techniques for Noise Robust Speech Recognition*. PhD thesis, Cambridge University, 1995.

[49] M. J. F. Gales, "Transformation smoothing for speaker and environmental adaptation," in *Proceedings of Eurospeech*, pp. 2067–2070, Rhodes, Greece, 1997.

[50] M. J. F. Gales, "Maximum likelihood linear transformations for HMM-based speech recognition," *Computer Speech and Language*, vol. 12, pp. 75–98, 1998.

[51] M. J. F. Gales, "Semi-tied covariance matrices for hidden Markov models," *IEEE Transactions on Speech and Audio Processing*, vol. 7, no. 3, pp. 272–281, 1999.

[52] M. J. F. Gales, "Cluster adaptive training of hidden Markov models," *IEEE Transactions on Speech and Audio Processing*, vol. 8, pp. 417–428, 2000.

[53] M. J. F. Gales, "Maximum likelihood multiple subspace projections for hidden Markov models," *IEEE Transactions on Speech and Audio Processing*, vol. 10, no. 2, pp. 37–47, 2002.

[54] M. J. F. Gales, "Discriminative models for speech recognition," in *ITA Workshop*, University San Diego, USA, February 2007.

[55] M. J. F. Gales, B. Jia, X. Liu, K. C. Sim, P. Woodland, and K. Yu, "Development of the CUHTK 2004 RT04 Mandarin conversational telephone speech transcription system," in *Proceedings of ICASSP*, Philadelphia, PA, 2005.

[56] M. J. F. Gales, D. Y. Kim, P. C. Woodland, D. Mrva, R. Sinha, and S. E. Tranter, "Progress in the CU-HTK broadcast news transcription system," *IEEE Transactions on Speech and Audio Processing*, vol. 14, no. 5, September 2006.

[57] M. J. F. Gales and S. J. Young, "Cepstral parameter compensation for HMM recognition in noise," *Speech Communication*, vol. 12, no. 3, pp. 231–239, 1993.

[58] M. J. F. Gales and S. J. Young, "Robust speech recognition in additive and convolutional noise using parallel model combination," *Computer Speech and Language*, vol. 9, no. 4, pp. 289–308, 1995.

[59] M. J. F. Gales and S. J. Young, "Robust continuous speech recognition using parallel model combination," *IEEE Transactions on Speech and Audio Processing*, vol. 4, no. 5, pp. 352–359, 1996.

[60] Y. Gao, M. Padmanabhan, and M. Picheny, "Speaker adaptation based on pre-clustering training speakers," in *Proceedings of EuroSpeech*, pp. 2091–2094, Rhodes, Greece, 1997.

[61] J.-L. Gauvain and C.-H. Lee, "Maximum *a posteriori* estimation of multivariate Gaussian mixture observations of Markov chains," *IEEE Transactions on Speech and Audio Processing*, vol. 2, no. 2, pp. 291–298, 1994.

[62] J. J. Godfrey, E. C. Holliman, and J. McDaniel, "SWITCHBOARD," in *Proceedings of ICASSP*, vol. 1, pp. 517–520, 1992. San Francisco.

[63] V. Goel, S. Kumar, and B. Byrne, "Segmental minimum Bayes-risk ASR voting strategies," in *Proceedings of ICSLP*, Beijing, China, 2000.

[64] P. S. Gopalakrishnan, D. Kanevsky, A. Nadas, and D. Nahamoo, "A generalisation of the Baum algorithm to rational objective functions," in *Proceedings of ICASSP*, vol. 12, pp. 631–634, Glasgow, 1989.

[65] R. Gopinath, "Maximum likelihood modelling with Gaussian distributions for classification," in *Proceedings of ICASSP*, pp. II–661–II–664, Seattle, 1998.

[66] R. A. Gopinath, M. J. F. Gales, P. S. Gopalakrishnan, S. Balakrishnan-Aiyer, and M. A. Picheny, "Robust speech recognition in noise - performance of the IBM continuous speech recognizer on the ARPA noise spoke task," in *Proceedings of ARPA Workshop on Spoken Language System Technology*, Austin, TX, 1999.

[67] A. Gunawardana, M. Mahajan, A. Acero, and J. C. Platt, "Hidden conditional random fields for phone classification," in *Proceedings of Interspeech*, Lisbon, Portugal, September 2005.

[68] R. Haeb-Umbach and H. Ney, "Linear discriminant analysis for improved large vocabulary continuous speech recognition," in *Proceedings of ICASSP*, pp. 13–16, Tokyo, 1992.

[69] R. Haeb-Umbach and H. Ney, "Improvements in time-synchronous beam search for 10000-word continuous speech recognition," *IEEE Transactions on Speech and Audio Processing*, vol. 2, pp. 353–356, 1994.

[70] T. Hain, "Implicit pronunciation modelling in ASR," in *ISCA ITRW PMLA*, 2002.

[71] T. Hain, P. C. Woodland, T. R. Niesler, and E. W. D. Whittaker, "The 1998 HTK System for transcription of conversational telephone speech," in *Proceedings of ICASSP*, pp. 57–60, Phoenix, 1999.

[72] D. Hakkani-Tur, F. Bechet, G. Riccardi, and G. Tur, "Beyond ASR 1-best: Using word confusion networks in spoken language understanding ," *Computer Speech and Language*, vol. 20, no. 4, October 2006.

[73] T. J. Hazen and J. Glass, "A comparison of novel techniques for instantaneous speaker adaptation," in *Proceedings of Eurospeech*, pp. 2047–2050, 1997.

[74] H. Hermansky, "Perceptual linear predictive (PLP) analysis of speech," *Journal of Acoustical Society of America*, vol. 87, no. 4, pp. 1738–1752, 1990.

[75] H. Hermansky and N. Morgan, "RASTA processing of speech," *IEEE Transactions on Speech and Audio Processing*, vol. 2, no. 4, 1994.

[76] F. Jelinek, "A fast sequential decoding algorithm using a stack," *IBM Journal on Research and Development*, vol. 13, 1969.

[77] F. Jelinek, "Continuous speech recognition by statistical methods," *Proceedings of IEEE*, vol. 64, no. 4, pp. 532–556, 1976.

[78] F. Jelinek, "A discrete utterance recogniser," *Proceedings of IEEE*, vol. 73, no. 11, pp. 1616–1624, 1985.

[79] H. Jiang, X. Li, and X. Liu, "Large margin hidden Markov models for speech recognition," *IEEE Transactions on Audio, Speech and Language Processing*, vol. 14, no. 5, pp. 1584–1595, September 2006.

[80] B.-H. Juang, "On the hidden Markov model and dynamic time warping for speech recognition — A unified view," *AT and T Technical Journal*, vol. 63, no. 7, pp. 1213–1243, 1984.

[81] B.-H. Juang, "Maximum-likelihood estimation for mixture multivariate stochastic observations of Markov chains," *AT and T Technical Journal*, vol. 64, no. 6, pp. 1235–1249, 1985.

[82] B.-H. Juang and S. Katagiri, "Discriminative learning for minimum error classification," *IEEE Transactions on Signal Processing*, vol. 14, no. 4, 1992.

[83] B.-H. Juang, S. E. Levinson, and M. M. Sondhi, "Maximum likelihood estimation for multivariate mixture observations of Markov chains," *IEEE Transactions on Information Theory*, vol. 32, no. 2, pp. 307–309, 1986.

[84] J. Kaiser, B. Horvat, and Z. Kacic, "A novel loss function for the overall risk criterion based discriminative training of HMM models," in *Proceedings of ICSLP*, Beijing, China, 2000.

[85] S. M. Katz, "Estimation of probabilities from sparse data for the language model component of a speech recogniser," *IEEE Transactions on ASSP*, vol. 35, no. 3, pp. 400–401, 1987.

[86] T. Kemp and A. Waibel, "Unsupervised training of a speech recognizer: Recent experiments," in *Proceedings of EuroSpeech*, pp. 2725–2728, September 1999.

[87] D. Y. Kim, G. Evermann, T. Hain, D. Mrva, S. E. Tranter, L. Wang, and P. C. Woodland, "Recent advances in broadcast news transcription," in *Proceedings of IEEE ASRU Workshop*, (St. Thomas, U.S. Virgin Islands), pp. 105–110, November 2003.

[88] D. Y. Kim, S. Umesh, M. J. F. Gales, T. Hain, and P. Woodland, "Using VTLN for broadcast news transcription," in *Proceedings of ICSLP*, Jeju, Korea, 2004.

[89] N. G. Kingsbury and P. J. W. Rayner, "Digital filtering using logarithmic arithmetic," *Electronics Letters*, vol. 7, no. 2, pp. 56–58, 1971.

[90] R. Kneser and H. Ney, "Improved clustering techniques for class-based statistical language modelling," in *Proceedings of Eurospeech*, pp. 973–976, Berlin, 1993.

[91] R. Kuhn, L. Nguyen, J.-C. Junqua, L. Goldwasser, N. Niedzielski, S. Finke, K. Field, and M. Contolini, "Eigenvoices for speaker adaptation," in *Proceedings of ICSLP*, Sydney, 1998.

[92] N. Kumar and A. G. Andreou, "Heteroscedastic discriminant analysis and reduced rank HMMs for improved speech recognition," *Speech Communication*, vol. 26, pp. 283–297, 1998.

[93] H.-K. Kuo and Y. Gao, "Maximum entropy direct models for speech recognition," *IEEE Transactions on Audio Speech and Language Processing*, vol. 14, no. 3, pp. 873–881, 2006.

[94] L. Lamel and J.-L. Gauvain, "Lightly supervised and unsupervised acoustic model training," *Computer Speech and Language*, vol. 16, pp. 115–129, 2002.

[95] M. I. Layton and M. J. F. Gales, "Augmented statistical models for speech recognition," in *Proceedings of ICASSP*, Toulouse, 2006.

[96] L. Lee and R. C. Rose, "Speaker normalisation using efficient frequency warping procedures," in *Proceedings of ICASSP*, Atlanta, 1996.

[97] C. J. Leggetter and P. C. Woodland, "Maximum likelihood linear regression for speaker adaptation of continuous density hidden Markov models," *Computer Speech and Language*, vol. 9, no. 2, pp. 171–185, 1995.

[98] S. E. Levinson, "Continuously variable duration hidden Markov models for automatic speech recognition," *Computer Speech and Language*, vol. 1, pp. 29–45, 1986.

[99] S. E. Levinson, L. R. Rabiner, and M. M. Sondhi, "An introduction to the application of the theory of probabilistic functions of a Markov process to automatic speech recognition," *Bell Systems Technical Journal*, vol. 62, no. 4, pp. 1035–1074, 1983.

[100] J. Li, M. Siniscalchi, and C.-H. Lee, "Approximate test risk minimization through soft margin training," in *Proceedings of ICASSP*, Honolulu, USA, 2007.

[101] H. Liao, *Uncertainty Decoding For Noise Robust Speech Recognition*. PhD thesis, Cambridge University, 2007.

[102] H. Liao and M. J. F. Gales, "Joint uncertainty decoding for noise robust speech recognition," in *Proceedings of Interspeech*, pp. 3129–3132, Lisbon, Portugal, 2005.

[103] H. Liao and M. J. F. Gales, "Issues with uncertainty decoding for noise robust speech recognition," in *Proceedings of ICSLP*, Pittsburgh, PA, 2006.

[104] H. Liao and M. J. F. Gales, "Adaptive training with joint uncertainty decoding for robust recognition of noise data," in *Proceedings of ICASSP*, Honolulu, USA, 2007.

[105] R. P. Lippmann and B. A. Carlson, "Using missing feature theory to actively select features for robust speech recognition with interruptions, filtering and noise," in *Proceedings of Eurospeech*, pp. KN37–KN40, Rhodes, Greece, 1997.

[106] X. Liu and M. J. F. Gales, "Automatic model complexity control using marginalized discriminative growth functions," *IEEE Transactions on Audio, Speech and Language Processing*, vol. 15, no. 4, pp. 1414–1424, May 2007.

[107] P. Lockwood and J. Boudy, "Experiments with a nonlinear spectral subtractor (NSS), hidden Markov models and the projection, for robust speech recognition in cars," *Speech Communication*, vol. 11, no. 2, pp. 215–228, 1992.

[108] B. T. Lowerre, *The Harpy Speech Recognition System*. PhD thesis, Carnegie Mellon, 1976.

[109] X. Luo and F. Jelinek, "Probabilistic classification of HMM states for large vocabulary," in *Proceedings of ICASSP*, pp. 2044–2047, Phoenix, USA, 1999.

[110] J. Ma, S. Matsoukas, O. Kimball, and R. Schwartz, "Unsupervised training on large amount of broadcast news data," in *Proceedings of ICASSP*, pp. 1056–1059, Toulouse, May 2006.

[111] W. Macherey, L. Haferkamp, R. Schlüter, and H. Ney, "Investigations on error minimizing training criteria for discriminative training in automatic speech recognition," in *Proceedings of Interspeech*, Lisbon, Portugal, September 2005.

[112] B. Mak and R. Hsiao, "Kernel eigenspace-based MLLR adaptation," *IEEE Transactions Speech and Audio Processing*, March 2007.

[113] B. Mak, J. T. Kwok, and S. Ho, "Kernel eigenvoices speaker adaptation," *IEEE Transactions Speech and Audio Processing*, vol. 13, no. 5, pp. 984–992, September 2005.

[114] L. Mangu, E. Brill, and A. Stolcke, "Finding consensus among words: Lattice-based word error minimisation," *Computer Speech and Language*, vol. 14, no. 4, pp. 373–400, 2000.

[115] S. Martin, J. Liermann, and H. Ney, "Algorithms for bigram and trigram word clustering," in *Proceedings of Eurospeech*, vol. 2, pp. 1253–1256, Madrid, 1995.

[116] A. Matsoukas, T. Colthurst, F. Richardson, O. Kimball, C. Quillen, A. Solomonoff, and H. Gish, "The BBN 2001 English conversational speech system," in *Presentation at 2001 NIST Large Vocabulary Conversational Speech Workshop*, 2001.

[117] S. Matsoukas, J.-L. Gauvain, A. Adda, T. Colthurst, C. I. Kao, O. Kimball, L. Lamel, F. Lefevre, J. Z. Ma, J. Makhoul, L. Nguyen, R. Prasad, R. Schwartz, H. Schwenk, and B. Xiang, "Advances in transcription of broadcast news and conversational telephone speech within the combined EARS BBN/LIMSI system," *IEEE Transactions on Audio, Speech and Language Processing*, vol. 14, no. 5, pp. 1541–1556, September 2006.

[118] T. Matsui and S. Furui, "N-best unsupervised speaker adaptation for speech recognition," *Computer Speech and Language*, vol. 12, pp. 41–50, 1998.

[119] E. McDermott, T. J. Hazen, J. Le Roux, A. Nakamura, and K. Katagiri, "Discriminative training for large-vocabulary speech recognition using minimum classification error," *IEEE Transactions on Audio Speech and Language Processing*, vol. 15, no. 1, pp. 203–223, 2007.

[120] J. McDonough, W. Byrne, and X. Luo, "Speaker normalisation with all pass transforms," in *Proceedings of ICSLP*, Sydney, 1998.

[121] C. Meyer, "Utterance-level boosting of HMM speech recognisers," in *Proceedings of ICASSP*, Orlando, FL, 2002.

[122] M. Mohri, F. Pereira, and M. Riley, "Weighted finite state transducers in speech recognition," *Computer Speech and Language*, vol. 16, no. 1, pp. 69–88, 2002.

[123] P. J. Moreno, *Speech recognition in noisy environments*. PhD thesis, Carnegie Mellon University, 1996.

[124] P. J. Moreno, B. Raj, and R. Stern, "A vector Taylor series approach for environment-independent speech recognition," in *Proceedings of ICASSP*, pp. 733–736, Atlanta, 1996.

[125] A. Nadas, "A decision theoretic formulation of a training problem in speech recognition and a comparison of training by unconditional versus conditional

maximum likelihood," *IEEE Transactions on Acoustics Speech and Signal Processing*, vol. 31, no. 4, pp. 814–817, 1983.

[126] L. R. Neumeyer, A. Sankar, and V. V. Digalakis, "A comparative study of speaker adaptation techniques," in *Proceedings of Eurospeech*, pp. 1127–1130, Madrid, 1995.

[127] H. Ney, U. Essen, and R. Kneser, "On structuring probabilistic dependences in stochastic language modelling," *Computer Speech and Language*, vol. 8, no. 1, pp. 1–38, 1994.

[128] L. Nguyen and B. Xiang, "Light supervision in acoustic model training," in *Proceedings of ICASSP*, Montreal, Canada, March 2004.

[129] Y. Normandin, "An improved MMIE training algorithm for speaker independent, small vocabulary, continuous speech recognition," in *Proceedings of ICASSP*, Toronto, 1991.

[130] J. J. Odell, V. Valtchev, P. C. Woodland, and S. J. Young, "A one-pass decoder design for large vocabulary recognition," in *Proceedings of Human Language Technology Workshop*, pp. 405–410, Plainsboro NJ, Morgan Kaufman Publishers Inc., 1994.

[131] P. Olsen and R. Gopinath, "Modelling inverse covariance matrices by basis expansion," in *Proceedings of ICSLP*, Denver, CO, 2002.

[132] S. Ortmanns, H. Ney, and X. Aubert, "A word graph algorithm for large vocabulary continuous speech recognition," *Computer Speech and Language*, vol. 11, no. 1, pp. 43–72, 1997.

[133] M. Padmanabhan, G. Saon, and G. Zweig, "Lattice-based unsupervised MLLR for speaker adaptation," in *Proceedings of ITRW ASR2000: ASR Challenges for the New Millenium*, pp. 128–132, Paris, 2000.

[134] D. S. Pallet, J. G. Fiscus, J. Garofolo, A. Martin, and M. Przybocki, "1998 broadcast news benchmark test results: English and non-English word error rate performance measures," Tech. Rep., National Institute of Standards and Technology (NIST), 1998.

[135] D. B. Paul, "Algorithms for an optimal A* search and linearizing the search in the stack decoder," in *Proceedings of ICASSP*, pp. 693–996, Toronto, 1991.

[136] D. Povey, *Discriminative Training for Large Vocabulary Speech Recognition*. PhD thesis, Cambridge University, 2004.

[137] D. Povey, M. J. F. Gales, D. Y. Kim, and P. C. Woodland, "MMI-MAP and MPE-MAP for acoustic model adaptation," in *Proceedings of EuroSpeech*, Geneva, Switzerland, September 2003.

[138] D. Povey, B. Kingsbury, L. Mangu, G. Saon, H. Soltau, and G. Zweig, "fMPE: Discriminatively trained features for speech recognition," in *Proceedings of ICASSP*, Philadelphia, 2005.

[139] D. Povey and P. Woodland, "Minimum phone error and I-smoothing for improved discriminative training," in *Proceedings of ICASSP*, Orlando, FL, 2002.

[140] P. J. Price, W. Fisher, J. Bernstein, and D. S. Pallet, "The DARPA 1000-word Resource Management database for continuous speech recognition," *Proceedings of ICASSP*, vol. 1, pp. 651–654, New York, 1988.

[141] L. R. Rabiner, "A tutorial on hidden Markov models and selected applications in speech recognition," *Proceedings of IEEE*, vol. 77, no. 2, pp. 257–286, 1989.

[142] L. R. Rabiner, B.-H. Juang, S. E. Levinson, and M. M. Sondhi, "Recognition of isolated digits using HMMs with continuous mixture densities," *AT and T Technical Journal*, vol. 64, no. 6, pp. 1211–1233, 1985.

[143] B. Raj and R. Stern, "Missing feature approaches in speech recognition," *IEEE Signal Processing Magazine*, vol. 22, no. 5, pp. 101–116, 2005.

[144] F. Richardson, M. Ostendorf, and J. R. Rohlicek, "Lattice-based search strategies for large vocabulary recognition," in *Proceedings of ICASSP*, vol. 1, pp. 576–579, Detroit, 1995.

[145] B. Roark, M. Saraclar, and M. Collins, "Discriminative N-gram language modelling," *Computer Speech and Language*, vol. 21, no. 2, 2007.

[146] A. V. Rosti and M. Gales, "Factor analysed hidden Markov models for speech recognition," *Computer Speech and Language*, vol. 18, no. 2, pp. 181–200, 2004.

[147] M. J. Russell and R. K. Moore, "Explicit modelling of state occupancy in hidden Markov models for automatic speech recognition," in *Proceedings of ICASSP*, pp. 5–8, Tampa, FL, 1985.

[148] G. Saon, A. Dharanipragada, and D. Povey, "Feature space Gaussianization," in *Proceedings of ICASSP*, Montreal, Canada, 2004.

[149] G. Saon, M. Padmanabhan, R. Gopinath, and S. Chen, "Maximum likelihood discriminant feature spaces," in *Proceedings of ICASSP*, Istanbul, 2000.

[150] G. Saon, D. Povey, and G. Zweig, "CTS decoding improvements at IBM," in *EARS STT workshop*, St. Thomas, U.S. Virgin Islands, December 2003.

[151] L. K. Saul and M. G. Rahim, "Maximum likelihood and minimum classification error factor analysis for automatic speech recognition," *IEEE Transactions on Speech and Audio Processing*, vol. 8, pp. 115–125, 2000.

[152] R. Schluter, B. Muller, F. Wessel, and H. Ney, "Interdependence of language models and discriminative training," in *Proceedings of IEEE ASRU Workshop*, pp. 119–122, Keystone, CO, 1999.

[153] R. Schwartz and Y.-L. Chow, "A comparison of several approximate algorithms for finding multiple (N-best) sentence hypotheses," in *Proceedings of ICASSP*, pp. 701–704, Toronto, 1991.

[154] H. Schwenk, "Using boosting to improve a hybrid HMM/Neural-Network speech recogniser," in *Proceedings of ICASSP*, Phoenix, 1999.

[155] M. Seltzer, B. Raj, and R. Stern, "A Bayesian framework for spectrographic mask estimation for missing feature speech recognition," *Speech Communication*, vol. 43, no. 4, pp. 379–393, 2004.

[156] F. Sha and L. K. Saul, "Large margin Gaussian mixture modelling for automatic speech recognition," in *Advances in Neural Information Processing Systems*, pp. 1249–1256, 2007.

[157] K. Shinoda and C. H. Lee, "Structural MAP speaker adaptation using hierarchical priors," in *Proceedings of ASRU'97*, Santa Barbara, 1997.

[158] K. C. Sim and M. J. F. Gales, "Basis superposition precision matrix models for large vocabulary continuous speech recognition," in *Proceedings of ICASSP*, Montreal, Canada, 2004.

[159] K. C. Sim and M. J. F. Gales, "Discriminative semi-parametric trajectory models for speech recognition," *Computer Speech and Language*, vol. 21, pp. 669–687, 2007.

[160] R. Sinha, M. J. F. Gales, D. Y. Kim, X. Liu, K. C. Sim, and P. C. Woodland, "The CU-HTK Mandarin broadcast news transcription system," in *Proceedings of ICASSP*, Toulouse, 2006.

[161] R. Sinha, S. E. Tranter, M. J. F. Gales, and P. C. Woodland, "The Cambridge University March 2005 speaker diarisation system," in *Proceedings of InterSpeech*, Lisbon, Portugal, September 2005.

[162] O. Siohan, B. Ramabhadran, and B. Kingsbury, "Constructing ensembles of ASR systems using randomized decision trees," in *Proceedings of ICASSP*, Philadelphia, 2005.

[163] H. Soltau, B. Kingsbury, L. Mangu, D. Povey, G. Saon, and G. Zweig, "The IBM 2004 conversational telephony system for rich transcription," in *Proceedings of ICASSP*, Philadelphia, PA, 2005.

[164] S. Srinivasan and D. Wang, "Transforming binary uncertainties for robust speech recognition," *IEEE Transactions of Audio, Speech and Language Processing*, vol. 15, no. 7, pp. 2130–2140, 2007.

[165] A. Stolcke, E. Brill, and M. Weintraub, "Explicit word error minimization in N-Best list rescoring," in *Proceedings of EuroSpeech*, Rhodes, Greece, 1997.

[166] B. Tasker, *Learning Structured Prediction Models: A Large Margin Approach*. PhD thesis, Stanford University, 2004.

[167] H. Thompson, "Best-first enumeration of paths through a lattice — An active chart parsing solution," *Computer Speech and Language*, vol. 4, no. 3, pp. 263–274, 1990.

[168] K. Tokuda, H. Zen, and A. W. Black, *Text to speech synthesis: New paradigms and advances*. Chapter HMM-Based Approach to Multilingual Speech Synthesis, Prentice Hall, 2004.

[169] K. Tokuda, H. Zen, and T. Kitamura, "Reformulating the HMM as a trajectory model," in *Proceedings Beyond HMM Workshop*, Tokyo, December 2004.

[170] S. E. Tranter and D. A. Reynolds, "An overview of automatic speaker diarisation systems," *IEEE Transactions Speech and Audio Processing*, vol. 14, no. 5, September 2006.

[171] S. Tsakalidis, V. Doumpiotis, and W. J. Byrne, "Discriminative linear transforms for feature normalisation and speaker adaptation in HMM estimation," *IEEE Transactions on Speech and Audio Processing*, vol. 13, no. 3, pp. 367–376, 2005.

[172] L. F. Uebel and P. C. Woodland, "Improvements in linear transform based speaker adaptation," in *Proceedings of ICASSP*, pp. 49–52, Seattle, 2001.

[173] V. Valtchev, J. Odell, P. C. Woodland, and S. J. Young, "A novel decoder design for large vocabulary recognition," in *Proceedings of ICSLP*, Yokohama, Japan, 1994.

[174] V. Valtchev, J. J. Odell, P. C. Woodland, and S. J. Young, "MMIE training of large vocabulary recognition systems," *Speech Communication*, vol. 22, pp. 303–314, 1997.

[175] V. Vanhoucke and A. Sankar, "Mixtures of inverse covariances," in *Proceedings of ICASSP*, Montreal, Canada, 2003.

[176] A. P. Varga and R. K. Moore, "Hidden Markov model decomposition of speech and noise," in *Proceedings of ICASSP*, pp. 845–848, Albuqerque, 1990.

[177] A. J. Viterbi, "Error bounds for convolutional codes and asymptotically optimum decoding algorithm," *IEEE Transactions on Information Theory*, vol. 13, pp. 260–269, 1982.

[178] F. Wallhof, D. Willett, and G. Rigoll, "Frame-discriminative and confidence-driven adaptation for LVCSR," in *Proceedings of ICASSP*, pp. 1835–1838, Istanbul, 2000.

[179] L. Wang, M. J. F. Gales, and P. C. Woodland, "Unsupervised training for Mandarin broadcast news and conversation transcription," in *Proceedings of ICASSP*, Honolulu, USA, 2007.

[180] L. Wang and P. Woodland, "Discriminative adaptive training using the MPE criterion," in *Proceedings of ASRU*, St Thomas, U.S. Virgin Islands, 2003.

[181] C. J. Wellekens, "Explicit time correlation in hidden Markov models for speech recognition," in *Proceedings of ICASSP*, pp. 384–386, Dallas, TX, 1987.

[182] P. Woodland and D. Povey, "Large scale discriminative training of hidden Markov models for speech recognition," *Computer Speech and Language*, vol. 16, pp. 25–47, 2002.

[183] P. Woodland, D. Pye, and M. J. F. Gales, "Iterative unsupervised adaptation using maximum likelihood linear regression," in *Proceedings of ICSLP'96*, pp. 1133–1136, Philadelphia, 1996.

[184] P. C. Woodland, "Hidden Markov models using vector linear predictors and discriminative output distributions," in *Proceedings of ICASSP*, San Francisco, 1992.

[185] P. C. Woodland, M. J. F. Gales, D. Pye, and S. J. Young, "The development of the 1996 HTK broadcast news transcription system," in *Proceedings of Human Language Technology Workshop*, 1997.

[186] Y.-J. Wu and R.-H. Wang, "Minimum generation error training for HMM-based speech synthesis," in *Proceedings of ICASSP*, Toulouse, 2006.

[187] S. J. Young, "Generating multiple solutions from connected word DP recognition algorithms," in *Proceedings of IOA Autumn Conference*, vol. 6, pp. 351–354, 1984.

[188] S. J. Young and L. L. Chase, "Speech recognition evaluation: A review of the US CSR and LVCSR programmes," *Computer Speech and Language*, vol. 12, no. 4, pp. 263–279, 1998.

[189] S. J. Young, G. Evermann, M. J. F. Gales, T. Hain, D. Kershaw, X. Liu, G. Moore, J. Odell, D. Ollason, D. Povey, V. Valtchev, and P. C. Woodland, *The HTK Book (for HTK Version 3.4)*. University of Cambridge, http://htk.eng.cam.ac.uk, December 2006.

[190] S. J. Young, J. J. Odell, and P. C. Woodland, "Tree-based state tying for high accuracy acoustic modelling," in *Proceedings of Human Language Technology Workshop*, pp. 307–312, Plainsboro NJ, Morgan Kaufman Publishers Inc, 1994.

[191] S. J. Young, N. H. Russell, and J. H. S. Thornton, "Token passing: A simple conceptual model for connected speech recognition systems," Tech. Rep. CUED/F-INFENG/TR38, Cambridge University Engineering Department, 1989.

[192] S. J. Young, N. H. Russell, and J. H. S. Thornton, "The use of syntax and multiple alternatives in the VODIS voice operated database inquiry system," *Computer Speech and Language*, vol. 5, no. 1, pp. 65–80, 1991.

[193] K. Yu and M. J. F. Gales, "Discriminative cluster adaptive training," *IEEE Transactions on Speech and Audio Processing*, vol. 14, no. 5, pp. 1694–1703, 2006.

[194] K. Yu and M. J. F. Gales, "Bayesian adaptive inference and adaptive training," *IEEE Transactions Speech and Audio Processing*, vol. 15, no. 6, pp. 1932–1943, August 2007.

[195] K. Yu, M. J. F. Gales, and P. C. Woodland, "Unsupervised training with directed manual transcription for recognising Mandarin broadcast audio," in *Proceedings of InterSpeech*, Antwerp, 2007.

[196] H. Zen, K. Tokuda, and T. Kitamura, "A Viterbi algorithm for a trajectory model derived from HMM with explicit relationship between static and dynamic features," in *Proceedings of ICASSP*, Montreal, Canada, 2004.

[197] B. Zhang, S. Matsoukas, and R. Schwartz, "Discriminatively trained region dependent feature transforms for speech recognition," in *Proceedings of ICASSP*, Toulouse, 2006.

[198] J. Zheng and A. Stolcke, "Improved discriminative training using phone lattices," in *Proceedings of InterSpeech*, Lisbon, Portugal, September 2005.

[199] Q. Zhu, Y. Chen, and N. Morgan, "On using MLP features in LVCSR," in *Proceedings of ICSLP*, Jeju, Korea, 2004.

[200] G. Zweig, *Speech Recognition with Dynamic Bayesian Networks*. PhD thesis, University of California, Berkeley, 1998.

[201] G. Zweig and M. Padmanabhan, "Boosting Gaussian Mixtures in an LVCSR System," in *Proceedings of ICASSP*, Istanbul, 2000.